Fine Homebuilding® *on*

Doors and Windows

Fine Homebuilding® *on* Doors and Windows

The Taunton Press

Cover photo by John Lively

...by fellow enthusiasts

First printing: August 1990
International Standard Book Number: 0-942391-56-X
Library of Congress Catalog Card Number: 89-40589
Printed in the United States of America

A FINE HOMEBUILDING Book

The Taunton Press, Inc.
63 South Main Street
Box 5506
Newtown, Connecticut 06470-5506
U.S.A.

CONTENTS

INTRODUCTION

Doors, windows, and skylights do more than provide access, ventilation, and light. The way they're made and detailed adds considerably to the quality of the spaces they serve. In some cases they are the dominant architectural features of the house.

The 32 articles in this book were collected from back issues of *Fine Homebuilding* magazine,* and deal with many aspects of building and installing doors and windows. If you're looking for a way to flush-mount a skylight over your stairwell, to hang a houseful of doors or to build a greenhouse addition, this book is a good place to start looking for ideas and finding expert solutions to construction problems. In addition to learning how to trim doors and windows, you'll also find here a number of articles on building and rebuilding traditional millwork items, such as fanlights, entries and multiple-pane sash.

—John Lively, editor

*The six volumes in the *Fine Homebuilding* on... series are taken from *Fine Homebuilding* magazine numbers 1 through 55, 1981 through mid-1989. A footnote with each article tells when it was originally published. Product availability, suppliers' addresses, and prices may have changed since then.

The other five titles in the *Fine Homebuilding* on... series are *Builder's Tools; Frame Carpentry; Floors, Walls and Stairs; Foundations and Masonry;* and *Baths and Kitchens.* These books are abridgements of the hardcover *Fine Homebuilding* Builder's Library.

Connecticut River Valley Entrance

Reproducing a famous 18th-century doorway

by Gregory Schipa

This past spring we were moving into the last stages of the restoration of the Burley house, a beautiful Georgian structure built in Newmarket, N.H., around 1710. It had been dismantled, moved to Riverton, Vt., and reerected on a new site.

Although we had found a Federal entrance on the building at its original site, framing mortises in the structure left no doubt that the house had once boasted double, 24-in. wide doors. As part of the restoration, the owner wanted us to recreate a Connecticut River Valley broken-pediment entrance. Since these beautiful entrances became fashionable around 1750 in the Newmarket area, she surmised that the old Burley house might have gotten one as part of a facelift to reflect the family's increasing affluence. The proportions of the building were right, and I was easily persuaded. After all, a job like this is a joiner's dream. We decided that it would be best to join such an entry by hand in the original manner. We had only to do a little research and then find our favorite example.

Development of the style—In 1636, settlers of the Massachusetts Bay Colony began to move westward from Boston to the Connecticut River Valley. Preferring the isolation and dangers of the wilderness to the overcrowding and religious intolerance in the Boston area, they first settled Springfield, Mass., and then Hartford, Windsor and Wethersfield, Conn. The migration continued for another 100 years, extending up the Connecticut River to New Hampshire and Vermont. It was not long before these pioneers began adding distinctive touches to the houses they built.

Most of the early work in the Colonies had been Jacobean in style, and was executed by English craftsmen, but the Connecticut River Valley towns soon developed their own unique details and embellishments, while staying with the same basic house plan. With the start of the Georgian Period, change accelerated. It can most easily be traced in the treatment of the front entrance, which rapidly evolved from a simply framed batten door through embellishment with the common architrave, transom and crown to the introduction (about 1700) of pilasters, double doors and the classic three horizontal members: architrave, frieze and cornice. Connecticut River Valley craftsmen lavished ever-increasing creativity on these entrances, apparently in celebration of the lessening dangers and growing wealth and freedom of the period. Soon cornice members evolved into full pediments, and before the middle of the 18th century, these elaborate pediments were further embellished by being broken and ended with volutes or rosettes.

Jim Eaton

The historic Burley house was moved from its original location to central Vermont, and its new entry was modeled after the one on the Rev. John Williams house in Deerfield, Mass.

Finding our model—Some of the very best examples of the Connecticut River Valley entrance can be found at Old Deerfield Village, Mass. Perhaps the best known of these is on the Rev. John Williams house (see p. 11). We chose to reproduce this particular entrance because its door opening was the same size as the Burley house rough opening. Also, the dimensions of the two houses were almost the same, and the window placements were similar. The Williams entrance also seemed to represent the best of the Connecticut River Valley pilastered doorways, with an early Jacobean feeling reflected in its steep, handcarved moldings (very few of which are classical), and its high raised-panel tombstone pedestals. Its massive broken-scroll pediment is very steep and abrupt, and its double pilasters are deeply fluted, with carved rosettes in the necking. It is perfectly balanced and well detailed. We hoped we could match the work of the 18th-century craftsman who built it.

Getting started—My foreman, Richard Tintle, and I first went to Deerfield to measure, draw and photograph the entryway. We figured it would take the two of us at least four weeks to do the job. We built the entrance primarily of 5/4 and ¾-in. clear local white pine. The stock had to be clear because we were cutting all moldings by hand. We could usually get two moldings from each 1x4 board, but cost was very much a factor, so we tried to use every extra strip of pine for the smaller moldings.

Very little was necessary to prepare the building to receive the work. The size of the rough opening was already dictated by the double stud mortises in the frame, and at 51 in. wide by 80 in. high, it was almost exactly the same as the one in the Williams house. The windows in the Burley house were far apart, and we had sheathed with ⅝-in. plywood, so we had a large, flat, blank space on which we drew the outline of the entire entrance. This gave us an immediate feel for its impressive dimensions, and made it easy for us to figure how much wood we would need for our first step, the underlayment. We also rigged a tarp so the entry would be under cover until it was completed and primed.

The underlayment, more properly called the clapboard catch, consists of the boards to which the entire built-up entrance detail is attached. We used two 8¾-in. wide boards of nominal 1-in. stock on each side, planning to hide the seam under the pilasters. Two horizontal 11½-in. wide boards across the top extended above the eventual level of the drip cap (the horizontal member of the cornice), and high enough to accept the bottom of the broken-scroll pediment. Two more 1x12s were later used to back the pediment and match its curve. All the boards used for the clapboard catch were joined by hand and jackplaned. The Williams entrance had been further embellished

From *Fine Homebuilding* magazine (October 1982) 11:36-41

with quoin work (carved V-grooves imitating masonry joints). On our project, we carved these grooves with a V-gouge.

Pedestals and pilasters—Once we were done with the clapboard catch, we began working from the ground up. The base of a pilaster is the plinth block. Because it is close to the moist ground, it is usually the first part of an entrance to rot out. This had happened on the Williams house, where the original plinth blocks had been replaced very shabbily with old rough-cut studs. We used solid sugar-pine blocks on the Burley house, hand-dressed to 7 in. high, 12 in. wide and 2¾ in. deep.

The pedestals for the pilasters rise immediately from the plinth blocks. They are doubled, the under-pedestal being 2½ in. wider than the one on top. They both called for carved quoin work, and the surface pedestal on each side is a tombstone panel (photo right). We were surprised to observe that on the Williams house these little panels were not joined in the customary manner, with stiles and rails, but were actually carved 1-in. boards. We followed suit on our bench, using butt chisels and a V-gouge.

The double pilasters themselves represented quite a challenge. Both the pilasters and under-pilasters were fluted, and their entasis, or taper, had to be reflected in the fluting. To do this, we gradually raised the plane to reduce the depth of each flute toward the top. This also had the effect of reducing its width of cut. We found the Stanley #55 multiplane (photo below right) handy for this job, but its fence could not be used. We had to fabricate one that would follow the taper. The final 16 in. or so of each flute had to be carved with a U-gouge after every few passes, or the plane would ride up. I also clamped a small tab of sheet metal under the stop-block, extending from it to the end of each flute. This protected the flat area beyond the flutes from being scarred as the plane's bottom passed over it.

The fluted part of the pilasters is about 56 in. long, but the pilasters themselves extend to about 77 in., past the capital molding group and carved necking, under the architrave and frieze, finally ending at the bottom of the cornice, where the molding of dentiled corona steps around its top. The pilasters taper from 7$\frac{7}{16}$ in. at the bottom to 6½ in. at the top. The under-pilasters are proportionately wider, reflecting the same shadow-line as the pedestals. Pilaster and under-pilaster each project 1$\frac{1}{16}$ in., and they are both fluted on their sides as well as on their fronts. Before finishing the pilasters, we laid out and carved the 6-in. dia., six-petal rosette on the necking of each. When we finished this part of the project, we had already worked 70 man-hours.

The moldings of the pedestal cap and those of the pilaster base form a single, solid group. They are quite primitive, mostly a series of hollows and rounds—early Jacobean rather than classical. Nevertheless, they have considerable impact because of the depth to which they're cut and the distance they project. We cut out these and other moldings with antique planes (photo top right) and the Stanley #55. We found

The tombstone panel of one of the pilaster pedestals is nailed up, below. The under-pedestal extends 1¼ in. out from each side of the face panel, and both sit on the solid sugar-pine plinth block. Two 8¾-in. boards make up the underlayment, or clapboard catch, to which everything else is attached. All the elements have been gouged to resemble stonework.

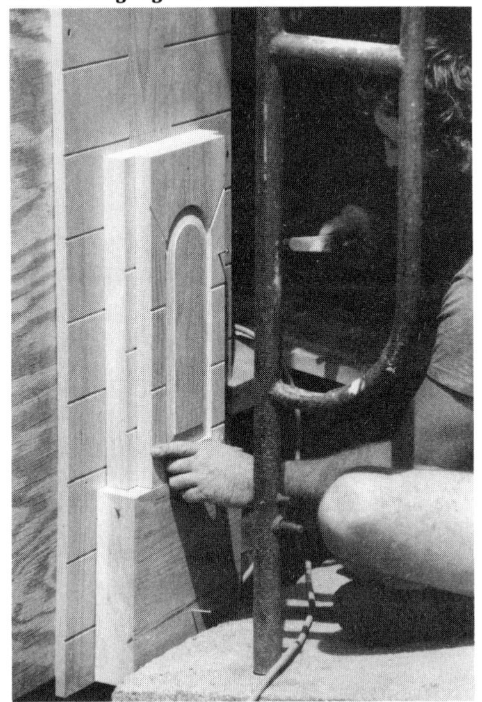

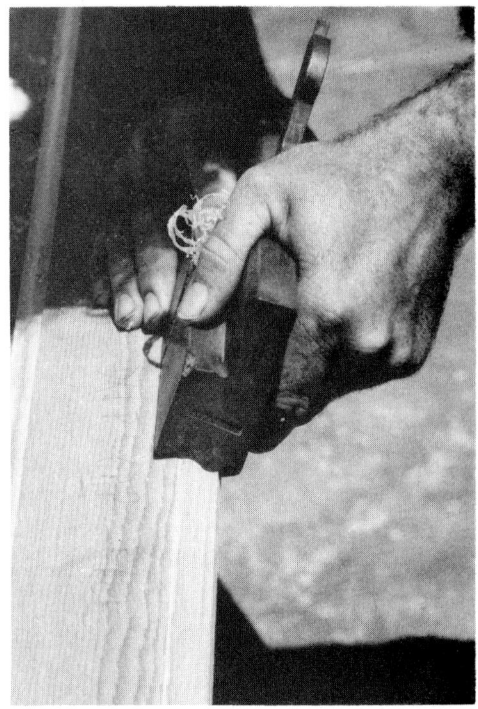

Antique planes, like the one above, are profiled to shape a specific molding. They came in handy on this job, where so many traditional shapes were required. A Stanley #55 plane (below) was used to cut the pilasters' fluting and, as shown here, to shape many of the entryway's moldings. This tool, designed for joiners, accepts blades of many different types and patterns.

Photos, except where noted: Gregory Schipa

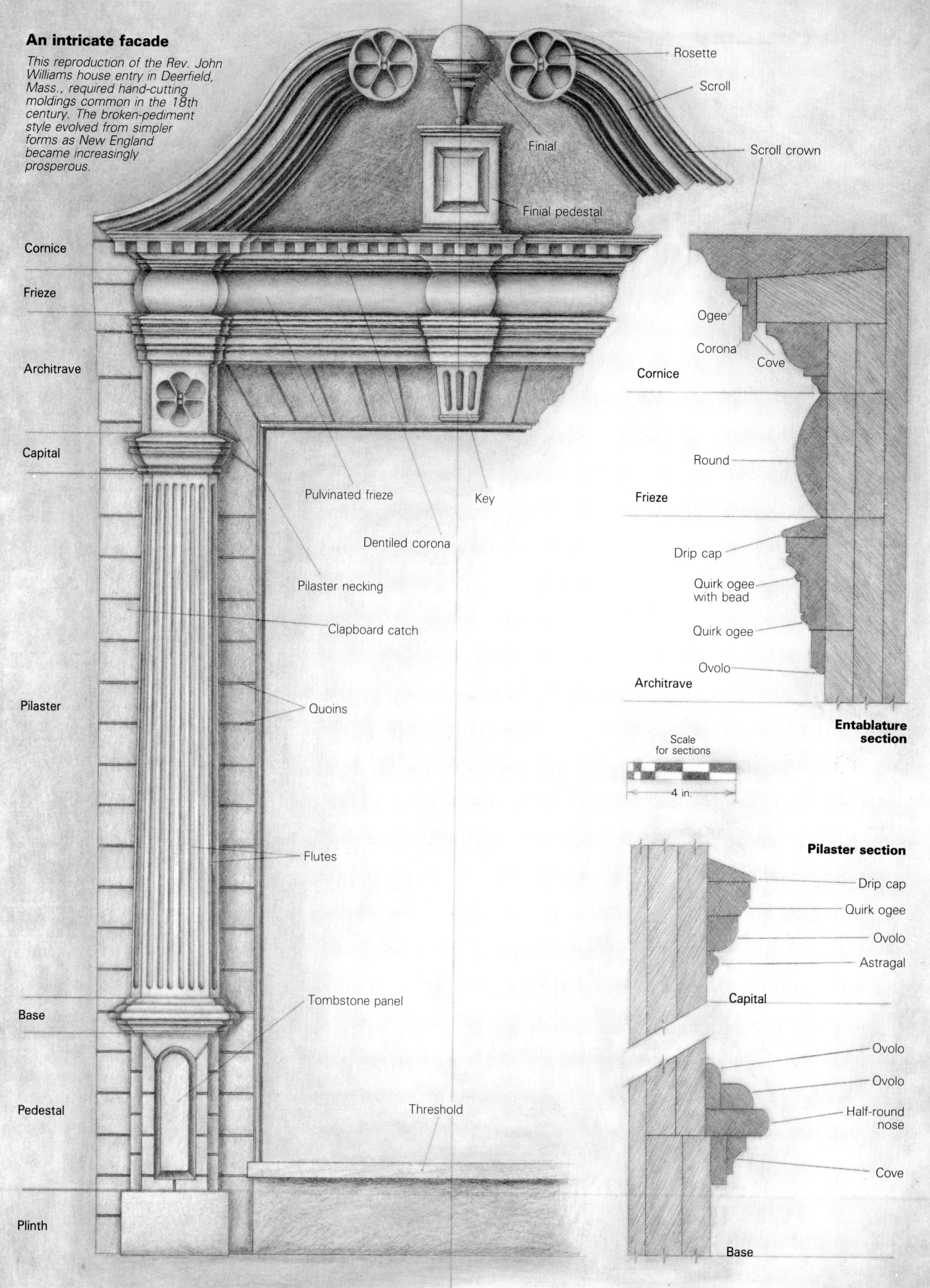

Illustration: E. Marino III

The perilous history of the Rev. John Williams and his house

In histories of Deerfield, Mass., the Rev. John Williams is known as "the redeemed captive." His house was burned to the ground and several of his children were killed during the French-fomented Indian raid of 1704, and Williams was carried off a prisoner to Canada. His wife was murdered on the march, and his 10-year-old daughter Eunice was adopted by the Caughnawagas. She converted to Catholicism and eventually married an Indian named Amrusus—a process that one partisan chronicler called "her lapse into barbarism." In 1706, Williams himself was exchanged for a Frenchman held prisoner by the English. His loyal congregation enticed him back to the wilderness from the safety of Boston by building him a new house in 1707.

When John Williams died, the house passed first to his second wife, and then, upon her death in the 1750s, to their son Elijah. A subsequent owner of Elijah's house bequeathed the structure to Deerfield Academy in 1875.

In the late 19th century, old houses were often destroyed with little regard for their historic value, and the school planned to raze the house to make room for a new building. But George Sheldon, president of the Pocumtuck Valley Association, saved the Williams house—and its glorious entry—by mounting a campaign that linked it closely with the Rev. Mr. Williams, the most romantic figure in the town's history. As researchers Amelia F. Miller and Donald L. Bunce put it, "With skillful omissions and careful wording, he allowed his readers to believe that this house was the one built in 1707." It wasn't. Miller and Bunce say that the 1707 house was pulled down, and a new house built in 1759 or 1760. Elijah had become a man of means and wanted a grander residence.

Miller and Bunce also found evidence that the great broken-pediment entryway was built by Lieutenant Samuel Partridge in the summer of 1760. Partridge had been with General Wolfe at Quebec, and was related to Elijah by marriage. He must also have been a master of his trade who could demand top dollar. Elijah's account book for September shows a disbursement of £39/0-2 (39 pounds, two pence) in an era when journeyman carpenters were making only a few shillings a day.

—Mark Alvarez

that applying molding around a double pedestal and pilaster requires three times the miter work as on single elements.

The pilaster capital molding group of the Williams entrance lines up just below the head jamb. It is steep and simple, which is appropriate to the period, but its moldings are more recognizably classic, with a ⅞-in. astragal, a 1¼-in. ovolo and a ⅝-in. ogee. The group has its own beveled drip cap, cut from a ¾-in. board and shaped with a bench plane.

The architrave—We thought we had done a few miters up to this point, but as we worked on the 95 pieces of the architrave band of moldings, the word began to take on a new meaning. The band has five components (including a spacer), and it steps around both double pilasters and the tapered key. The key, with its own handcarved flutes, extends from the door head to the bottom of the cornice, and matches the size and taper of the quoins. The architrave band is 5 9/16 in. wide, made up of a 1 9/16-in. bottom piece cut to a ¼-in. ovolo molding, followed by a 1¾-in. piece cut to a ⅝-in. ogee, and then a 1½-in. ogee with a bead. The full band has its own ¾-in. tapered drip cap, like the ones on the pilaster capitals. The architrave is ½ in. deep at its bottom and 1⅛ in. at its top—considerably less projection than the pilaster base and cap moldings.

Immediately above the architrave, and sitting flush upon its drip cap, the craftsman at the Williams house had fashioned a rounded, or pulvinated, frieze. We found that we could fabricate this element with a large hollow, a tool that was specifically made for the pulvinated molding, or with a bench plane, scratch-stock and sandpaper. The inaccuracy inherent in the handmade process could easily be worked out with a little additional sanding after the frieze had been mitered and fitted.

Sitting upon the frieze of the Williams entrance, and fitted against the bottom of the cornice like a bed molding, we found the classic 18th-century crown molding. Consisting of a small ogee on the bottom, with a bead above and a large reverse ogee on top, this molding

The mitered moldings of the entry required painstaking sanding. At left, Richard Tintle works on the crown molding beneath the cornice. The clapboard catch, grooved to resemble quoins, is visible to the left of the pilaster.

The scrolls (bottom left) were glued up from sugar-pine planks, and the curves were smoothed with an antique compass plane. On the facing page, the completed pediment.

The finial between the scrolls sits on a paneled pedestal, built as a sturdy box. The side panels are flat, but the front panel, below, is carved. The box will get a hipped top.

(photo above left) was one of 18th-century New England's most consistently popular architectural embellishments. We nailed it in place before we put up the frieze. We were very lucky to own the appropriate antique plane, because its use was the key to cornice and pediment detail on this particular entrance. The crown molding is repeated on the broken scroll, where it is mitered into the small ogee on the cornice fascia.

The pediment—A hundred and twelve man-hours after leaving to measure the famous entrance in Old Deerfield, we were ready to start the pediment. We had made and fitted 270 pieces up to this point. But the craftsman who worked on the original had saved his most creative efforts for last. The massive cornice, rising in steep reverse curves to form the broken pediment, and terminating in bold carved rosettes, is one of the most impressive details of 18th-century domestic architecture. We knew it would be a challenge to reproduce.

The cornice structure begins with the usual horizontal element, which serves as the drip cap for the entrance below. On our entry, we reproduced this with a 2-in. thick, 5-in. wide sugar-pine plank, which we beveled with a jack plane to shed water. We added 1-in. blocks to simulate the step-outs of the pilasters and the key below. The fascia applied to this plank has a $\frac{5}{16}$-in. reveal on the bottom, and a small $\frac{1}{4}$-in. cove cut into its back below that. Its front bottom, also called the corona, has small dentil-

Jim Eaton

like cuts, 1¼ in. square. These can be cut with a fillister plane, or a rabbet plane with a fence clamped to it. It is then topped with a ⅝-in. ogee molding, which we mitered into the bottom ogee of the crown molding, on the return of the broken-scroll pediment to the house.

We laid out the scrolls using a string and a pencil on a very large table. This was hard because we had only our photographs as guides, and we wanted to match the proportions of the original builders. To miss here would sacrifice a lot of very careful work. Once we felt we had proper proportions, we drew out templates on plywood, including all of the cove and thumbnail bed moldings that would follow the reverse curve pediment immediately below it. Then, as we cut them off, each snake-like strip became a different template. The scrolls themselves we cut from laminated sugar-pine planks. We used a beautiful old compass plane to smooth them after an initial rough cutting (photo facing page, bottom). On the rough scroll we attached a ½-in. fascia (again with the reveal and cove on the bottom), but here we discovered a real quirk of the craftsman who built the Williams entry. He had used a different crown on the scroll than on its returns to the house. Although he used a large reverse ogee with a bead and a small cove as his crown molding cut on the scroll, he used the very typical large reverse ogee, bead and small ogee for the returns. The large reverse ogees were the same size, however, and were mitered at their meeting. The small ogee was mitered into the ogee above the corona. It was a tricky resolution. We made the curved moldings by simply carving them and then working them hard with scratch-stock and sandpaper. We used the large reverse curve of our old crown molding plane as the template for the profile of the scroll crown.

The resolution—Large carved rosettes resolved the scroll tops. They were very similar to the rosettes on the pilaster necking, but were 2 in. larger, and our carving was in bas relief. We did all the rosette carving with U-shaped gouges. The pediment, from clapboard catch to the outside edge of the crown molding, is the same 8-in. depth as the block the rosettes are carved in, so we could terminate all moldings on the circular barrel of the rosette.

Between the rosettes, the pediment on the Williams house had been further embellished with a turned finial on a carved, paneled pedestal. Here we again struggled with our layout, for the finial had been too high on the real entrance for us to measure. Our drawing had to be our template, and when we finally settled on the proportions, we turned it out easily on the lathe—just as it had been done originally, if with a different power source. We left a large round tenon to fit into a mortise in the roof of the pedestal. The pedestal itself we made by first creating a box 8 in. wide by 12 in. high by 4 in. deep (small photo, facing page). Then we hand-carved a raised panel on its front, and flat panels on its sides (another oddity of the Williams entry). We fashioned a hipped top, flat only on the small portion meeting the bottom of the finial. A deep mortise accepted the finial's substantial tenon. We flashed the cornice, then set the pedestal and finial in place.

So there we were. Three-hundred and fourteen handcarved pieces, 154 man-hours of labor, and we had ourselves an entrance—and a lot of respect for those Connecticut River Valley craftsmen. In hopes of its lasting half as long as the original, we used handmade rosehead nails where they would show, and galvanized nails everywhere else. We also used almost two quarts of glue and untold pounds of lead for flashing. In addition to the horizontal cornice, we also flashed the scroll pediment, overlapping the edge of the crown molding by almost ¼ in. We even fitted the finial with a small leaden skull-cap.

After the awesome entrance, we built the doors, almost as an afterthought. They were real beauties, though. The client chose as our model a fine set of early four-paneled doors, again from Old Deerfield. They were fitted with beaded vertical battens on the inside, large strap hinges, and a huge bar and staples lock. The battens were fastened with clinched roseheads, the stiles and rails with pegs. The client was lucky enough still to have the huge granite step from the original house. We had only to move it gently up under the plinth blocks to make the job complete. □

Gregory Schipa is president of Weather Hill Restoration Co. in Waitsfield, Vt.

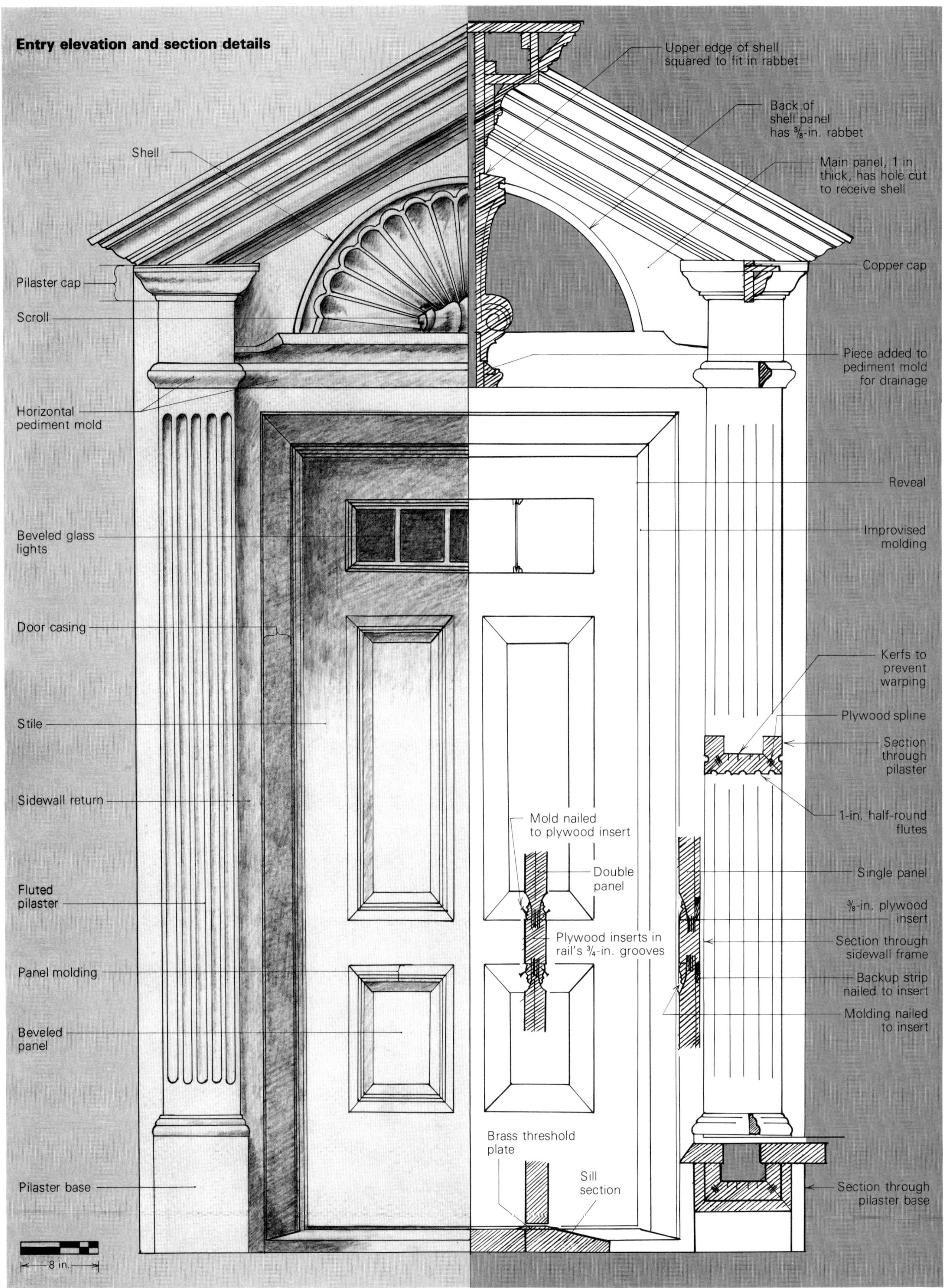

Illustrations: Roz Lapidus

From *Fine Homebuilding* magazine (June 1981) 3:14-18

Formal Entryway

A challenging adaptation of traditional designs

by Sam Bush

It was my pleasure to have been working for Pat Dillon, of Canby, Ore., when Mr. and Mrs. Alfred Herman asked him to build a new entry for their home in Portland. The 18-year-old existing door was not particularly well designed or constructed, and the Hermans wanted to replace it with one that incorporated various traditional elements. Pat often said that the doorway is the most important part of a residence because it introduces the people who live there. Pat and I liked the Hermans, and wanted them to be introduced properly.

We were a good combination of clients and carpenters for work of this sort. Pat, of Canby, Ore., is an experienced contractor with high standards for everything he undertakes. My own background is primarily in furniture, cabinetmaking and woodcarving, but I enjoy challenging finish work. Pat had earned the confidence of the Hermans before I'd come along. It was encouraging to know that they trusted us to do a good job and that our workmanship would be appreciated.

Design and planning—Helen Herman had a pretty clear idea of how she wanted the entry to look: pedimented, with fluted pilasters on either side, and paneled sidewalls. The basic concept was simple elegance. So Pat and I did our research, visiting, photographing and sketching many doorways. This was great fun for me, as I had studied Colonial and Federal work all over New England. We presented our ideas, and after many conversations with Helen we arrived at the proportions and lines that captured what she had in mind. The final design is a personal one, though it is based on traditional elements. We consulted on every part, right down to the panel moldings, which were created with shaper cutters profiled to our drawings. Relying on the good eyes of three people, we strove for balance and attractiveness throughout.

There were two stages of construction: making all the parts in the shop (which took six weeks), and installing them (which took seven days). First came shop drawings and calculations. Pat had made discreet investigations of the substructure of the existing entry without disrupting its appearance too much. We used his measurements for planning and ordering lumber.

For everything but the shell, we used clear, vertical-grain old-stand fir, a wood readily available in the Northwest. It's expensive, but great to work with, except for a tendency to splinter. (One afternoon, as I was working to extract one of a long line of slivers, Pat jokingly accused me of trying to take home wood from the job.)

The frames—Our first job was making the door stiles and rails. All of it was made of 2¼-in. stock, and we chose the pieces very carefully to be sure they would stay straight after installation. All the joints were deep mortise and tenon, haunched top and bottom for strength, and glued. There was also a ¾-in. wide groove run up the inside of every framing member in the lower part of the

A blend of New England architectural elements, the new entryway features custom moldings, a shell carving and a copper-clad roof.

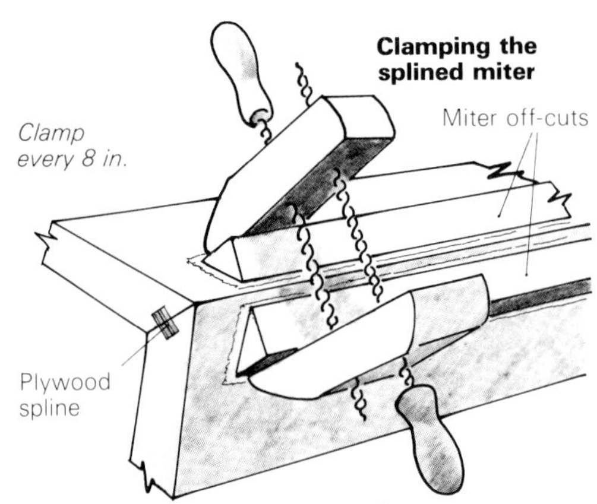

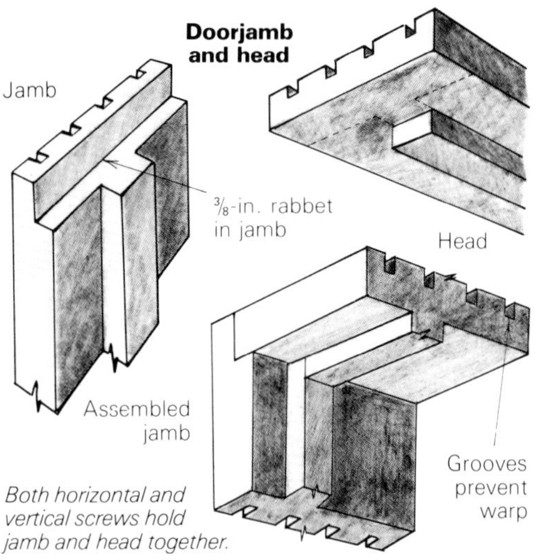

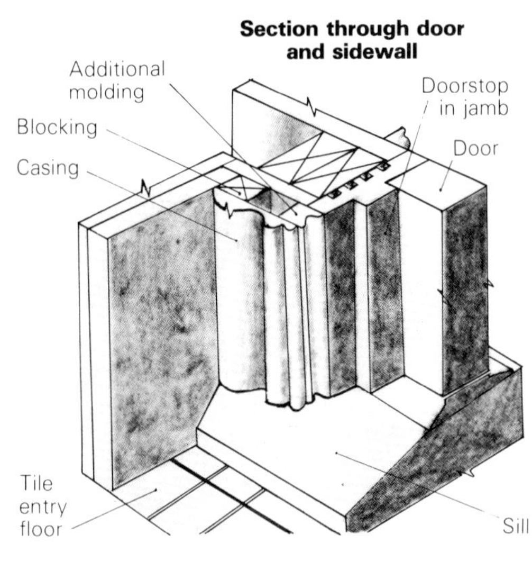

Door-frame construction is haunched mortise and tenon; grooves (photo below) will receive plywood inserts for attaching panel moldings. Long ears on door sill (photo bottom) will support door casing. Molding to receive lower right panels is already in place. Other joinery details are shown in the drawings at left.

door, to hold the plywood inserts to which panel moldings would be attached. We didn't put a groove in the top panel of the door, since a beveled-glass window would be located there.

Next came the framing for the ceiling and sidewalls, which was made from 1½-in stock, with mortise-and-tenon construction like the door. We ran ⅜-in. grooves for plywood inserts that would be arranged slightly differently from those in the door. These frames were assembled with a wider stile on the inside so that when the door casing was installed, overlapping it, the remainder showing would be the same size as the outside piece. We also built mitered returns on sidewall and ceiling frames where they would turn and lie on the face sheathing of the building. We decided early on to use as few nails as possible, knowing their tendency to move and show through, so we joined these miters with ¼-in. plywood splines and glued them with temporary clamp blocks made from the miter offcuts, as shown in the drawing, top left.

Panels and molding—Door panels could have been made of single 2¼-in. pieces, but we chose to use two thicknesses back-to-back so that they could adjust individually to indoor and outdoor conditions, avoiding the tendency to warp. Each 1⅛-in. thickness was made of two edge-glued boards; we ran the edges on the shaper with a special glue-joint cutter to prevent creep along the joint. Almost every joint in the job was glued with Weldwood resorcinol waterproof glue.

We shaped the panels with a ⅜-in. tongue to correspond with the plywood inserts. With the other cutter, we ran the molding that would hold the panels in place without inhibiting their natural movement. It was fairly easy to miter the molding on the chop saw and install the panels, which had been treated with wood preservative and primer paint. (All the parts were treated on both sides with two coats of Houston's #3 preservative and one coat of Kelley-Moore exterior primer before final installation.) We used Titebond to glue the molding to the plywood inserts before setting two finish nails on the horizontals and three on the verticals.

The other custom molding for this job was the 4½-in. wide door casing, similar to the panel molding, but larger. We were able to make some of it with shaper and router cutters, but the main line was really handmade, with a few table-saw cuts and a lot of scraping and planing.

Jamb, cornice and pilasters—We used solid stock for the door jamb, and grooved it on the back to prevent possible warp. The sill was screwed to the rabbeted jamb, flat under the door and sloping 5° on the outside, with long ears left on either side for the casings to rest on (photo left). After the jamb and sill were glued and screwed, the door was dimensioned and beveled, leaving a 3/32-in. space all around to allow for paint. Hanging involved setting three brass-plated ball-bearing butts, 4½ in. by 4½ in., using a Stanley door guide and a router to make the hinge mortises. We also fitted the brass latch and the Baldwin Lexington-pattern lock hardware, and I can tell you it's an exacting task to get the precise mechanism just right and working smoothly. (And this is no time to drill a hole in the wrong place.) The striker plate was not set into the jamb until later. Making up and nailing on the molding that holds in the beveled glass window completed the door work.

Next, we drew a full-sized plan, on plywood, of one-half the triangular frieze and cornice area. We figured out the length of the roof and the sizes of the box structure to which the cornice would be attached, the main panel, the shell and its moldings. With this information we were able to make up all these parts and get them waterproofed and primed. Since the roof boards over the cornice and the pilaster caps were to be sheathed in copper, they were sent out to the sheet-metal shop for fitting.

The last step before installing the entryway was making the pilasters, which were mitered and splined like the sidewall returns. The major difficulty here was the fluting, which—like everything else—had to be just right: five 1-in. diameter half-round grooves on the face, with exactly ½ in. between them, and one groove on each side surface. I made these on the shaper, relying on a sample block to make the settings. Since the fluting didn't extend to the end of the pilasters, I had to drop the work onto the cutter and pull it off for each cut in accordance with layout lines on the work and the fence. I ran the cuts from both edges. The middle one was quite a reach for Pat's shaper, but everything worked out. The cutter left a peculiar-looking shape where it came out of the work, so I finished the ends of each flute with my carving gouges. After this we cut the pilasters to length and mounted a three-part molding at the top of each. We were finally done in the shop.

Installing the door—While we were doing all this, the tile men had been at the Hermans' laying a 4-in. by 8-in. quarry tile surface on the floor of the new entry. Our first task was to demolish the original finish work. The existing sidewalls were ½-in. plywood with molding nailed on to make them look paneled. Fitting all over was casual, and construction choices were ill considered. In fact, a great deal of water had gotten into the trim and rotted it.

Removing the old doorway was last, and meant we could start rebuilding. It also meant we would be there until the new door was in, since we weren't about to leave for the evening with a big hole in the house. Setting the jamb and sill assembly had to be particularly accurate to fit both the existing interior and the new exterior parts, but the installation was done conventionally. To help take the weight of this heavy door, we ran 3-in. #8 screws through the jamb into the rough framing. These were in the hinge mortises where they would be covered. With the door in place and swinging, I marked and mortised the jamb for the striker plate and dust box. I took special care here, wanting as a matter of pride to have the door close both easily and tightly.

Entry paneling and casing—After the door was installed, we started work on the ceiling unit, the panel of which was pierced to accommodate the overhead lamp. This frame had to be set an exact and uniform distance from the jamb so the door casing would fit evenly. We then leveled, shimmed and nailed it along its left and right edges with as few 16d galvanized finish nails as possible.

The sidewalls were next. Like the ceiling, they were located parallel to the jamb—no margin for error here. They were also scribed to fit both the ceiling (covering its nails) and the pitched tile floor, while showing as a plumb line on the outside face. To prevent rot, we held all our finish work ¼ in. off the tile.

It was at this point that we discovered the only serious mistake of the job. Somewhere back in the calculation stage, I had erred by an inch in the sidewall measurements, making them narrower than they should have been. While the casing fit to the walls, the visible vertical stiles were now of uneven widths, and the balanced effect was spoiled. They say the *real* skill of a carpenter shows in how he gets himself out of the mistakes he gets himself into. Pat earned his money that day, inventing a piece of molded stock that brought out the surface of the jamb and solved the problem. This disaster cost us time, materials and more than a few grey hairs, but as sometimes happens, the adjustment turned out to be an improvement. The additional molding looked natural and made the doorway look even better.

With sidewalls and improvised molding in, and corresponding blocking next to the wall, it was my job to set the handmade casing. There was a lot to keep in mind while I was working: There had to be an even reveal along the door and perfect fit to the wall; both of the miters had to be tight and meet the corners of the wall and ceiling frames; the side pieces had to be cut at 5° on the bottom to sit on the angled sill—and there was no extra stock.

Around the entry—Next, we moved to the face of the house. We determined where the exterior vertical sticks would rest, and cut the cedar siding to the chalk line using a portable circular saw against a straight board. Likewise we made the angled cuts for the peak, after carefully measuring to ensure that the peak was plumb above the ceiling midline. When the side pieces were laid, they were tight against the siding, with their surfaces in line with the sidewall returns. They were also bedded in a thick corner bead of butyl caulking, which sealed the joint. We later ran another tiny bead of colored caulking along the surface of the siding where it met the side pieces.

Next, we notched the roof boards to fit over the siding, and toenailed them into the sheathing. We screwed blocking under their outer edges, then fit them into their copper covers, working a long flange of the metal up behind the siding. We bent the covers around the roof front edge, where we discreetly nailed it into the blocking with four copper nails. An overlapping, interlocking joint between copper sheets along its apex rendered the roof watertight. Then we mounted the 1-in. thick main panel on shims that brought it out flush with the 1½-in. trim. The back of the shell carving would eventually fit into the space behind the panel, so the shims were placed accordingly.

With the background and roof in place, we fastened the pilasters plumb, leaving an equal width of vertical boards on both sides of each one. The top moldings were already attached, so we needed only to add the angled cap pieces with their matching copper covers. At the base we nailed on the 1-in. thick spline-mitered base assemblies, after first scribing them to fit the tile. We cut and installed the molding along the tops of the bases to complete the pilasters.

The cornice box went up next, against the roof and backwall blocking, mitered at the top. Putting up the detailed cornice moldings was just about the most trying part of the whole job. A combination of factors—difficult miters inside and outside, close reveals, scribing to the siding, no extra material—left no room for error. But patience and a sharp block plane prevailed, and the parts fit as they were meant to. We used needlepoints, 1¼-in. galvanized finish nails, to fasten these moldings. They hold tightly and don't split the thin molding edges. Driven flush with the surface, they hold paint and don't show. We then installed the two-part detail inside the cornice box and above the shell. The moldings were mitered at the top, and died out against the pilaster caps. The last of the molding work was the strong horizontal piece right above the entry. Compared to the others, this one was simple because it was level, with regular 45° miters. Both Pat and I were glad to be done up above. We had made an unbelievable number of trips up and down the ladders.

The installation was completed except for the shell. We agreed on the lines of its elliptical shape and put up a cardboard mockup to see how it looked, and to get Helen's approval. There had been talk of a fan light or other details above the door in the beginning, but it was the shell that really fit into Helen's idea of the entry. That made me feel pretty good, because it had been my suggestion.

The shell—Back at the shop we used resorcinol to glue up the block from three layers of clear ponderosa pine, which carves nicely. We paid close attention to getting truly flat surfaces, edge and face, laying the boards so their edges did not

To install the entry, the old siding was first removed, left. At right, the entryway complete but for the shell carving—the cardboard mock-up holds its place.

line up. The bottom piece was longer than the main area and was made with an angled top edge glued in an angled rabbet, so the ears on the edges of the shell would have a draining surface. Pat had the further good idea of adding another 1-in. piece to the scroll areas so it could project out beyond the rest. This we did, and it added greatly to the finished effect.

I did the carving at home on my basement bench in three-and-a-half days of work. The photos below illustrate the sequence of events. After carefully tracing the pattern onto the wood from the full-sized drawing with carbon paper, I drilled a series of holes of uniform depth very close to each other around the scroll area. This made it possible to remove the wood with chisel and mallet, working always from the outside to the middle and having the long chips break off against the scroll. When the drilled holes started to disappear, it was time to quit at the middle. Other than running my hand back and forth across the dish-shaped depression, feeling for bumps, there was no guide or template to measure the curve. The dark gluelines showing through were a big help in this respect. The bottom of this area was left angled about 12° for drainage.

The scroll itself was next, a careful sculpting that involved much checking and rechecking to make sure the two halves came out the same size with natural-looking spirals. I established the ends first and then shaped the rest of the block into a balanced, compound-curving mass that looked a bit like a football with its ends cut off. On this bulk, I laid out the lines for the center bump and the ribbon-like strips at the ends. The careful modeling of the whole block first made it relatively easy to carve in the molded shape and have it come out right. The ends, however, were just plain work. They were difficult to visualize because the spiral shape climbs upward to the tip, and does not lie in a flat plane.

With the scroll complete, I laid out an even spacing around the inside and extended lines from the perimeter to the center with a flexible straightedge. Guided by these sets of tapering lines, I carved the shell flutes with a long-bend gouge. Here again the lamination lines were helpful in establishing depth. The ridges between the flutes were the only part of the dished shape that remained, and I carved these with a slightly concave surface. Then, using a router and straight mortising bit, I reduced the area between the scallops and the edge to a depth of ⅜ in., trimming up the lines with carving tools.

The next move was bandsawing out the block's elliptical shape and the curving ears on each end. I did it with the table on a 12° angle to provide the needed pitch. At this time I also routed a 1-in. deep rim around the back, to allow weathertight mounting of the shell. I was then able to carve the deep concave molding of the base, and carefully sand the whole piece with papers ranging from 60 grit to 120 grit. After a thorough waterproofing during which the wood was lightly heated between coats to maximize absorption, the shell was primed.

Back at the Hermans, mounting involved cutting a hole to receive the shell in the background panel with a reciprocating saw. That allowed the shell's rim and base to lie on the surface of the background panel, so that its profile could be marked exactly. The shell overlapped the panel by 2 in. We routed out ⅜ in. from the 2-in. overlap with a hand-held router, trimming to the marked outline with chisels where necessary. This left a wide, flat rabbet into which the shell block fit. The bottom edge of its base was worked to correspond with the angled surface of the molding on which it sits. We ran two wide beads of butyl caulk into the rabbet, and permanently mounted the shell with ten large galvanized finish nails in predrilled holes. A thin bead of caulking along the joint between the shell and background panel completed the job.

While I was finishing the shell, Pat got the weatherstrippers to the job to install the brass threshold and brass interlocking trim all around the door. He also installed a copper mail chute the sheet-metal people had made, to carry mail from an extruded brass mail slot to a box located in a closet. This box collects the mail, and is fitted with a door so there isn't heat loss through the mail slot.

That completed our work, except for Pat's overseeing the subcontractors. The concrete men formed up and poured a new set of steps up the bank, and gave them an attractive exposed-aggregate surface. Then the wrought-iron people installed the new railings. And the painters came and finished our work with two coats of exterior oil-base white. For us it was a very successful and satisfying conclusion to a big job. The Hermans were certainly pleased, and said so, and that was our bonus. □

Sam Bush heads the wood program at the Oregon School of Arts and Crafts in Portland.

Making the shell begins with gluing up three layers of ponderosa pine, edges staggered for strength, top left. Small block at front is for scroll; holes will aid in carving flutes. Top right, fluting is laid out in dished-out shells; scroll carving is complete. With flutes carved, above left, semicircular groove is routed and block is bandsawn to shape, above right. A concave molding base completes the job.

Hanging an Exterior Door

From framing the rough opening to mortising for hinges, installing a door requires patience and precision

by Jared Emery

An exterior door has to fit its jamb tightly, though not so tightly that it causes a shoulder separation when you try to open it. You'll have few problems if you're using a prehung door—one that's already fitted to its jambs. But if your roughed-in opening is an odd size or you want to use your own special hinges, say, or you want jambs made out of something other than stock pine to match your casings or paneling, you may want to hang your own door. This can be an awkward and frustrating task, but getting the proper fit is largely a matter of cutting the door to a median dimension that will accommodate seasonal changes in the wood.

The rough opening for the exterior jamb should be framed in larger than the door by the total thickness of both legs of the jamb plus ½ in. on either side for shim space. For example, a 3-0/6-8 door with a jamb made from 1-in. stock would require a rough opening of 39 in. wide and about 82 in. high.

The legs and head of the jamb should be 1-in. thick. The sill should be of 6/4 stock. These units are available unassembled, prerouted with the sill planed and the stop in place. Or you can make your own.

First give the jambs a primer coat and then assemble the frame, using waterproof glue and three 12d casing nails through each side jamb into the head. To avoid splitting the wood, don't nail within about ½ in. of the jamb edges. Square up the frame and then nail the sill in place the same way. Finish by driving two 16d casing nails through each corner of the head and the sill into the jambs.

Next you have to prepare the rough opening

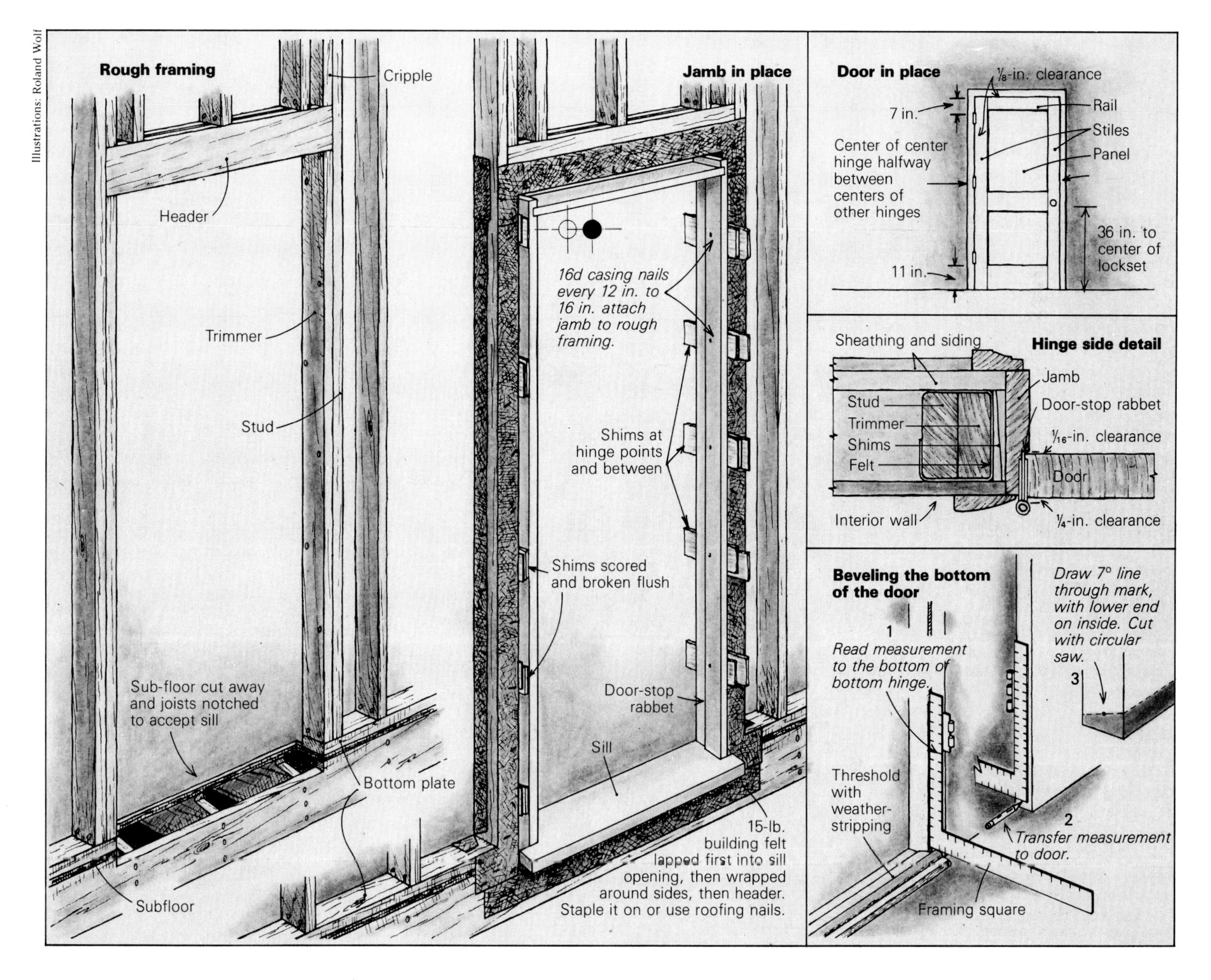

to receive the frame. Wrap the studs of the opening with 12-in. to 16-in. wide strips of 15-lb. roofing felt for extra insurance against water infiltration. The sill of the frame must finish out flush with the top of the finish floor. In most cases, you will have to cut out at least the subfloor, and you may have to notch or trim the rim joist as well.

Fitting the frame to the opening—The rest of the job depends on how well you set the frame. The two most common errors in this work are racking the frame in its opening, and plumbing one leg while leaving the head out of level. Once the frame is within the opening, level the head by shimming beneath the sill under the appropriate leg. It's best to use two wedges when you shim, driving one in from each side. Next, shim the hinge jamb plumb, setting double wedges at the hinge locations; drive 10d finishing nails only partway through the jamb into the trimmers to hold it snugly against the wedges. Plumb the other jamb the same way, add several more shims along each side to keep things steady, recheck for square, then drive and set 16d casing nails through the center of the jambs into the trimmers every 12 in. to 16 in. Then remove the 10d nails, which held things temporarily in place.

Up to this point, what you've done would apply equally to prehung doors or to those you would hang yourself. If you were working with a prehung door, you would already have had to determine which way you wanted your door to swing. If you're hanging your own, you could wait this long to decide whether the door will open on the right-hand or left-hand side. All exterior doors open to the inside. If there is a light switch on the wall, the door should swing away from it so someone entering can reach inside to turn on the lights.

One of the best kept secrets of carpentry is how to determine the "hand" of a door. This is important when you're ordering some lock hardware or a prehung unit. Simply stated, when the door is closed and you are standing on the inside, if the knob is on the right, it's a right-hand door.

Fitting the door to its jamb—This part of the job requires that you carefully measure the opening and know the tolerances demanded by the weatherstripping you intend to use. Of the many types available, the one I like most is the spring-type weatherstripping. Made of bronze-colored light-gauge aluminum, it has a nailing surface and a flange that angles back toward the outside, against which the door will close. If you use this material, you should size the door ¼ in. less than the jamb width. This dimension allows ⅛-in. gaps at both the hinge and lock sides. The optimim clearance for the lock stile is 3/16 in., but it's a good idea to leave the door a little wide at first, so that any irregularities in the jamb may be planed into the door after it has been hung.

At this point, you should do any trimming on the hinge side. This will leave a full stile for mortising the lock hardware. This is especially important if the door has a glass panel in it. Having a lock mortise break through the stile into a glass panel is guaranteed to be a heart-stopping experience.

Finally, plane a 5° bevel on the lock stile, so that the leading edge of the door will clear the angled portion of the weatherstripping. I use an electric Rockwell Portaplane; a well-sharpened jack plane works well, too.

The height of the door should be ¼ in. less than the total jamb height, for ⅛-in. clearance at the head and just enough clearance at the bottom to allow it to close. (The bottom of the door should be cut to fit its weatherstripped threshold later, after you hang the door.) Check the jamb again for squareness and plumb before you do any cutting.

After the door has been fit to the jamb, it is ready to be hung. Hinges should be selected according to the size of the door. Most exterior doors are solid core and 1¾ in. thick. Building codes require them to be at least 3 ft. wide. The best hinges to use to support this much weight are 4-in. by 4-in. loose-pin butt hinges.

Installing the hinges—Place the top of the top hinge 7 in. from the top of the door, the bottom of the bottom hinge 11 in. from the bottom of the door, and a third one centered between the two. If you're using a ¾-in. thick casing on the inside of the jamb, you should leave ¼ in. of the hinge leaf outside the mortise. This brings the outside of the hinge's barrel into alignment with the outer edge of the casing, and allows the door to swing 180° without running afoul of the casing and levering the hinges loose.

For laying out and mortising hinges, the door must be held securely with the hinge side up. A door buck is best for this (see the box below) though you can brace the door with a sawhorse by nailing a strip of wood to each. First mark out the hinge positions and then cut the mortises, as explained on the next page.

Next, mount the hinge leaves. Drill the pilot holes for the screws just slightly off center, toward the closed edge of the mortise. The wedging action of the screw head will pull the hinge tightly into the mortise.

When the hinge leaves are in place, the door can be set into its jamb, wedged up to the correct height and shimmed against the hinge side of the jamb. Using a utility knife, score the jamb slightly above and below each hinge. The knife-marks indicate the exact location of the hinge, something a pencil line cannot do.

It is good practice to make the mortises on the jamb 1/16 in. or so narrower than the mortises on the door. This is especially true if either the door or the jamb is to be painted. This difference in widths allows for the thickness of the paint between the door and the stop and for the expansion of the stop. Mortise the jamb the same way you mortised the door stile. But you'll find it more awkward to do precise work on a vertical surface than on a horizontal one.

Always put the pin in the top hinge first, so the door can hang while you catch your breath and ease your cramped arms and aching hands. If the mortises have been cut accurately, the hinge leaves will line up and fit together. If one of them doesn't, though, loosen the screws on both leaves of that hinge and tap them together with a hammer. Then insert the pin and retighten the screws.

There should now be ⅛-in. clearances on both the hinge and lock sides of the door. But

There are two common ways to hold a door while you work on one of its edges. Door bucks work well for me. You can buy these, but I make mine out of triangles of 2x8 stock, usually stair-stringer scraps. I nail two of these a door-width apart on a platform of ½-in. plywood long enough to hold the door. Then I fasten 2x4 blocks to each corner and in the middle as legs. The weight of the door deflects the plywood, pressing the tops of the triangles against the door. This holds it steady, and you can take it out with no fumbling around.

The other method is to set up the door parallel to a sawhorse and tack a small strip of wood to both the horse and the door with 4d nails. A single strip will hold the door, but this method is a bit more cumbersome than using a door buck, and the nailed strip gets in the way if you're planing. —*J.E.*

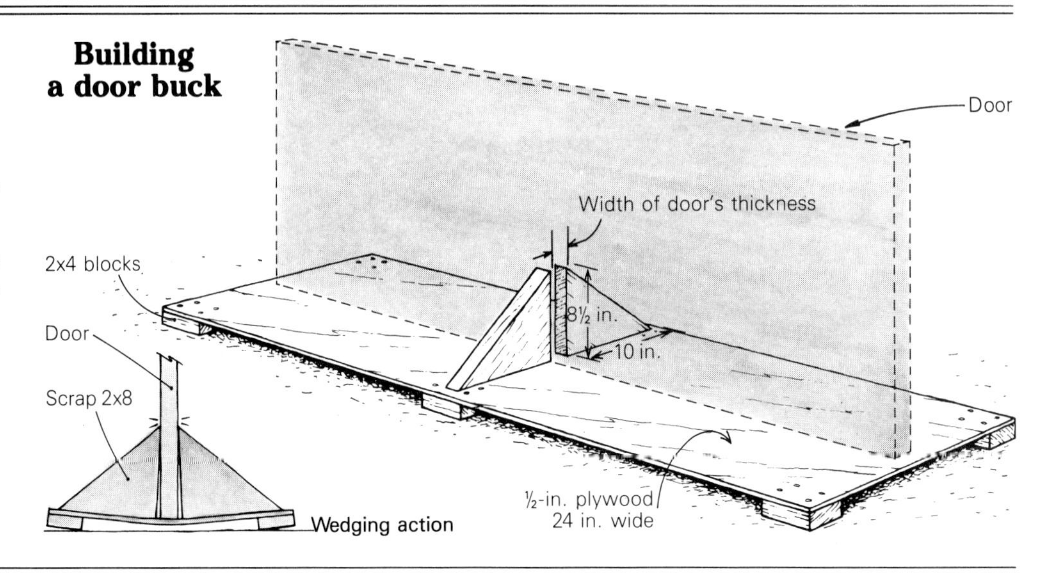

the optimum clearance on the lock side is $\frac{3}{16}$ in. Some carpenters like to trim the door to this tolerance before hanging it, but I've noticed that the general tendency is to trim off just a little too much. (You can shim out the hinges with cardboard in this situation, but that's a step I'd rather not have to take.) There are often slight irregularities in jambs that have to be accommodated, too. I prefer to check the door once it's hanging, scribe for the $\frac{3}{16}$-in. gap, then take it down again and plane to the line, maintaining the 5° bevel.

Hardware and weatherstripping—Next, let in the lock hardware and apply the weatherstripping. It is a good idea, when mortising for the lockset, to cut the mortise slightly deeper than it has to be. This will give you some room for adjustment if the door needs any planing in later years.

A high-quality lockset comes with an adjustable faceplate that can be set to the bevel on the door. Less expensive locksets don't, and setting the lock deeper into the door keeps the faceplates of moderately priced locksets from protruding at the lower side of the bevel.

Nail weatherstripping to the jamb with the flange angled toward the outside every 2 in., using the small-headed brads supplied. The spring flanges of bronze weatherstripping have to be mitered at the corners to fit, and it must be cut around the strike plate on the lock side of the jamb. I nail the flanges down above and below the strike to prevent excessive wear.

There are several satisfactory types of weatherstripping available for the threshold. The most common is an aluminum saddle with a rubber or vinyl insert. The saddle is notched to fit around the door stop, casing and jamb, so that the rubber insert will lie beneath the door. Once the saddle has been fitted and secured to the sill, the door must be cut to fit.

To find the cut-line across the bottom of the door, set a framing square against the rubber insert with the tongue of the square against the hinge-jamb leg. This will tell you if the saddle is lying square to the jamb, and will also give you a measurement from the rubber insert to the bottom of the bottom hinge.

With the door once again lying on the sawhorses, transfer this measurement to the center of the edge of the hinge stile. Then draw a 7° line through this mark, the lower end of the line being on the inside of the door. Where this line comes out on the inside, draw another line across the bottom face of the door. This is the cut-line. If the door has a veneer face, score it first with a knife to prevent chipping. Then set a circular saw at 7° and cut the door bravely. If you're going to paint the door, prime the cut edge at once. If you're going to stain it, do so at once and then apply varnish or one of the urethane finishes. Don't let a new door hang for long without a sealer or prime coat.

Rehang the door and check its fit. It should close snugly and give slight resistance when it is opened. □

Jared Emery is a writer disguised as a carpenter. He lives in Charlottesville, Va.

1

2

3

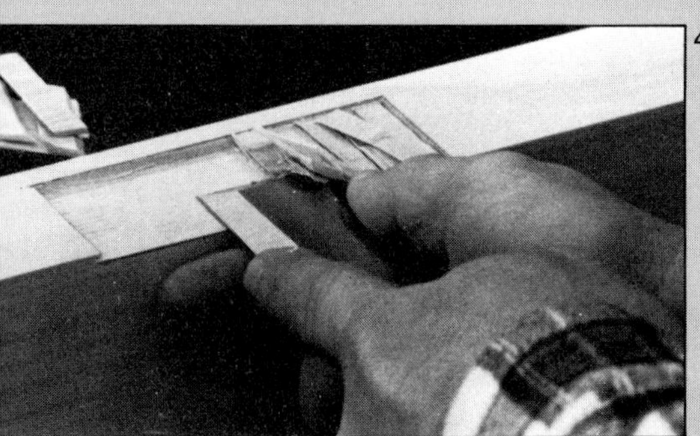
4

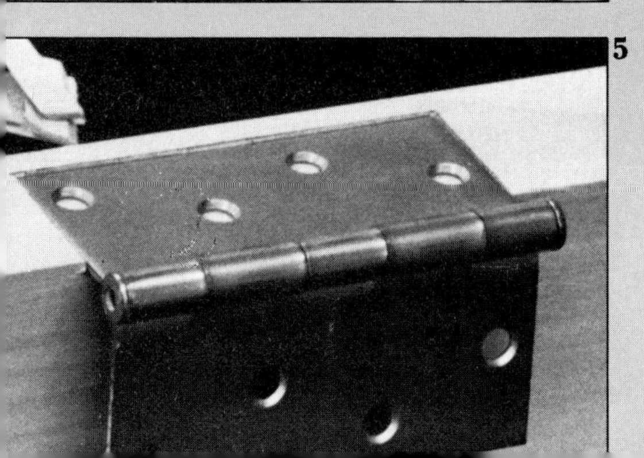
5

Mortising butt hinges

For mortising hinges, many tradesmen like to use a router. But with a good butt chisel, a hammer and a little practice, you can cut mortises for hinges on a single door as fast as you can with a router. Router setups take time, and you'll probably have to trim the cut with a chisel anyway. So why not do the whole job with a few hand tools?

Lay out the mortise first. The two lines across the grain must be struck with a knife to cut the wood fibers cleanly. The two lines with the grain—one for the back edge of the leaf, the other for the depth of the mortise—should be made with a scratch awl or other pointed tool. Don't use a pencil only, or you'll get imprecise results, and risk splintering the wood outside the area you want to excavate.

Determine the depth of the mortise by figuring how wide a gap you need between the door and the hinge jamb. For an interior door with no weatherstripping, you may want to cut the mortise slightly deeper than the leaf is thick, to make the gap neat and minimal. For an exterior door, the mortise will be as deep as the leaf is thick. The barrel of the hinge should be set out from the door enough so that when the door opens, it clears the casing.

With the layout done, score the wood in about $\frac{1}{4}$-in. wide increments down the length of the layout. Use a sharp chisel as wide as or wider than the mortise, and hold it at about 45°, as shown in photo 1. With a single blow from hammer or mallet, cut the wood tissue to within a hair of the depth line. About 10 of these cuts should do for a 4-in. butt. The closer you hold your chisel to 90°, the more force required to sever the wood fibers. Do not at this stage try to cut straight down on the lines at the ends of the mortise.

Next hold the chisel vertically, so its bevel is away from the line, and cut the back wall of the mortise with a few light taps of the hammer (photo 2). If you try to make this cut before scoring up the wood inside the mortise, you will probably split out the wood along the line, and ruin the whole job. This will happen because the still-intact waste will not yield and will force the chisel to act like a wedge instead of a cutting tool.

Having cut the back wall of the mortise, you should now pare away the scored tissue. Begin about $\frac{1}{2}$ in. away from one side, index the edge of the chisel (bevel up) in the depth line, and pare straight back to the rear wall (photo 3). This should not require much force, which means that you shouldn't risk running the chisel past the rear wall of the mortise. So be sure your chisel is sharp, and that you've scored the wood to the right depth before you begin horizontal paring.

The first cut should give you a nice flat surface the full width of your chisel. As shown in photo 4, you'll use this surface to register the chisel for subsequent cuts to get the whole mortise in the same plane. When you've pared away most of the waste, chisel the end walls square to the bottom. Again, don't try to make these vertical cuts until you've gotten the adjacent tissue out of the way.

When all's done, the hinge should fit snugly (photo 5). Another advantage in mortising with a chisel is that you can quickly make slight adjustments in the depth of the mortise, something that would take a lot of fiddling if you used a router. ***—John Lively***

Hanging Interior Doors

An organized approach to a demanding job that must be done right to be done at all

by Tom Law

Among carpenters, door hanging is a prestigious job. It requires skill, patience and a thorough understanding of the steps involved. I spent years hanging doors and fitting hardware when I was employed in commercial work. But for the last ten years I've been a general contractor, and hanging doors is still something I like to do. On those unhappy occasions when speed and economy are more important than quality, I'll install prehung doors, but even then prefer not to. Being able to hang doors properly gives you considerable flexibility in choosing the materials you use for door, jamb, stop and casing. And where custom materials are specified, you may have no other choice than to hang your own.

I generally hang doors in place, even though it costs more in terms of time and money. It gives me something constructive to do on site when I'm coordinating the various trades during the finishing stage. And my clients are assured of properly fitted doors that won't bind during wet weather and won't show large gaps between door and jamb during the winter.

On the average, it takes me about three hours to hang an interior door. This includes setting the jamb, fitting the door, mortising the door and jamb for hinges, hanging the door, applying the stop and casing and installing the lockset. If I have several doors to hang, I don't go through this sequence on each door; rather I move from door to door, repeating the same operation on each, until all are finished. More about this later.

Setting the jambs—Interior jambs are made of clear lumber. They need not be especially strong as the weight of the door is not borne by

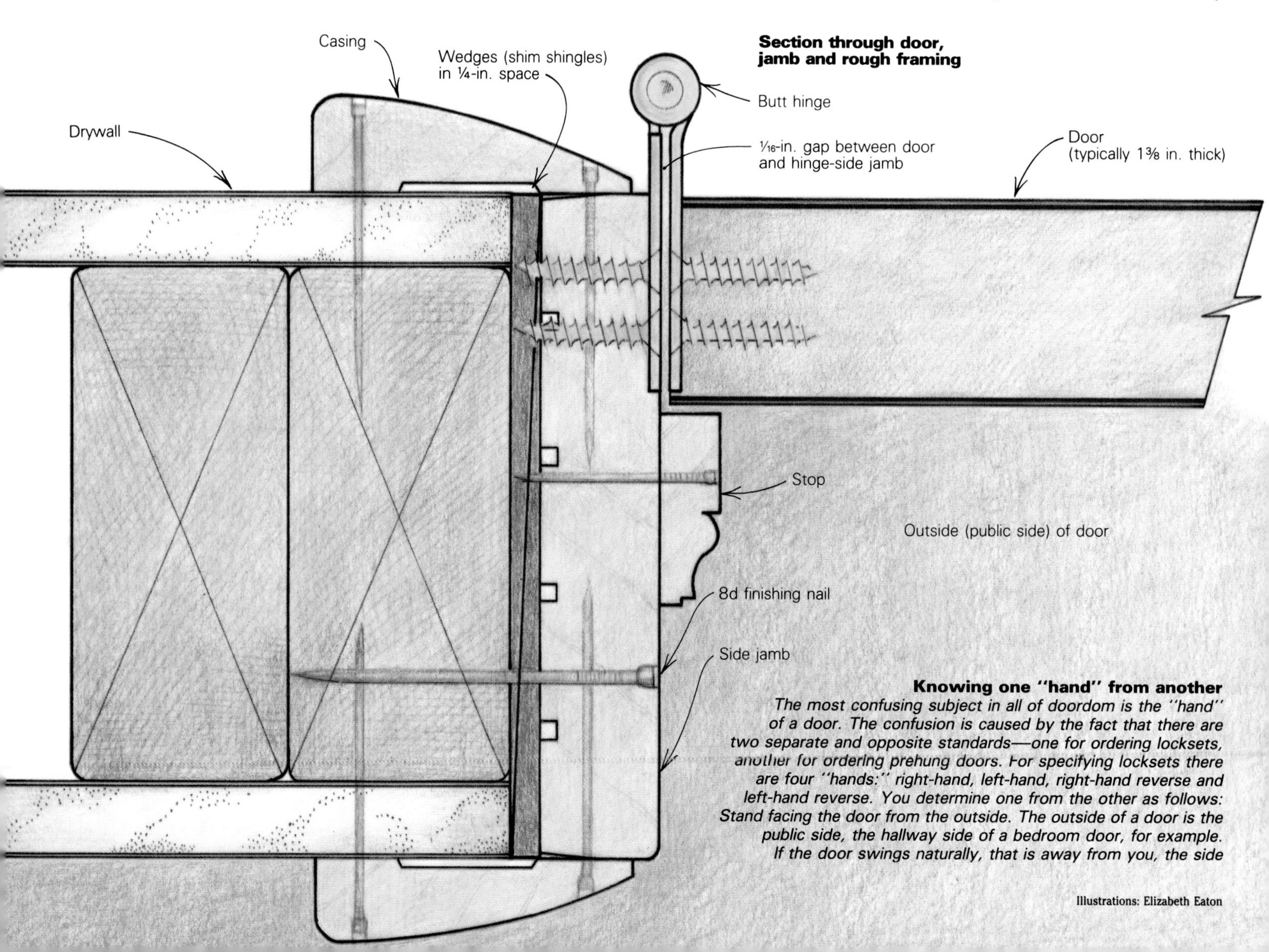

Knowing one "hand" from another

The most confusing subject in all of doordom is the "hand" of a door. The confusion is caused by the fact that there are two separate and opposite standards—one for ordering locksets, another for ordering prehung doors. For specifying locksets there are four "hands:" right-hand, left-hand, right-hand reverse and left-hand reverse. You determine one from the other as follows: Stand facing the door from the outside. The outside of a door is the public side, the hallway side of a bedroom door, for example. If the door swings naturally, that is away from you, the side

Illustrations: Elizabeth Eaton

the jamb but is transferred to the jack studs through the shim wedges (drawing, below). The wood should be straight-grained and free of warp. In the old days, the back side of the jamb was kerfed longitudinally to relieve any stress that might develop and cause the jamb to cup. These days, however, those kerfs have been reduced to shallow V-grooves, which will do little to control cupping and twist. Also, some new jambs are not continuous lengths of lumber end to end, but are made up of finger-joined pieces of clear pine. These jambs are acceptable if painted, but if you want to finish them naturally, you'll want to select clear, solid stock.

The jamb should be slightly wider than the thickness of the finished wall. For example, a nominal 2x4 stud wall with ½-in. drywall on each side should measure 4$\frac{9}{16}$ in. thick, and the standard jamb thickness for this wall is 4⅝ in. Walls are never perfect. They are either too thick or too thin, and it's a problem to make the door jamb and its casing fit perfectly. The head jamb is housed in dadoes cut into the side jambs, which means that an allowance must be made for the combined dado depth when cutting the head to length.

You should take great care when setting jambs to make them as accurate as possible and thereby eliminate unnecessary work when fitting the door. The first thing I do is make a spreader out of scrap ¾-in. plywood for the bottom of the jamb. The spreader's job is to hold the side jambs parallel, so its length should be exactly the distance between the side jambs at the head. It should be about 1 in. wider than the jamb.

After nailing the side jambs and head together and before standing the assembled jamb in place, I measure the rough opening. It ought to be about 2 in. larger all around than the door. This will leave about ¼ in. on each side of the jamb for shimming, assuming that the jambs are ¾ in. thick. A shim space this size is sufficient for plumbing the jamb in a rough opening, but it's a good idea to check the jack studs (also called *trimmers* or *cripples)* for plumb before setting the jamb. If they are out of plumb (more than ¼ in. top to bottom), you'll want to get them right. Next, stand the jamb in the opening, put the spreader in the bottom and temporarily shim between both sides at the top to hold it in place. (For shims, I use undercoursing-grade shingles and rip them on my table saw to about 1½ in. wide.)

Now try the head for level (photo next page, top left). If the finish floor will be carpet, the side jambs can rest directly on the subfloor. All you need to do is shim up under the low side to level the head jamb. But if the jamb has to sit on top of a finish floor like oak or tile, you begin by sawing off the bottoms of the side jambs in an amount that equals the thickness of your finish floor. Then set the jamb in the opening, block up the sides with a pair of scraps that equal the thickness of the finish floor, and shim up one side to get the head perfectly level. Now set your scriber to the distance of the shim space and scribe off the opposite leg, as shown in the photo next page, center left. Remove the jamb, trim off the scribe and replace the jamb with both legs on the blocks. The head is now level.

Place the spreader between the sides at the bottom, and again wedge the top firmly in place. The square ends of the spreader are used to bring the jamb sides into proper alignment. I temporarily wedge the bottom of the side jambs against the spreader. Next, I find the center of the head and mark it on the edge for a plumb line, drive in a 4d finish nail just to the side of the mark and hang a plumb bob on it; this way the center of the string coincides with the center of the pencil line. Then I draw a centerline across the face of the spreader. When hanging the bob, I adjust the line so that the point just misses the wood.

Plumbing and squaring the jamb is now just a matter of bringing the centerline on the spreader directly under the tip of the bob. Do this by loosening the shim shingle on one side of the jamb and tightening the one on the opposite side, as shown in the photo next page, bottom left. Next you should drive in the wedges so they are snug, but before nailing the jamb to the jack studs, lay a short straightedge against the wall and jamb edge to make sure that the jamb projects equally on both sides of the wall.

Now you can nail one side near the top with

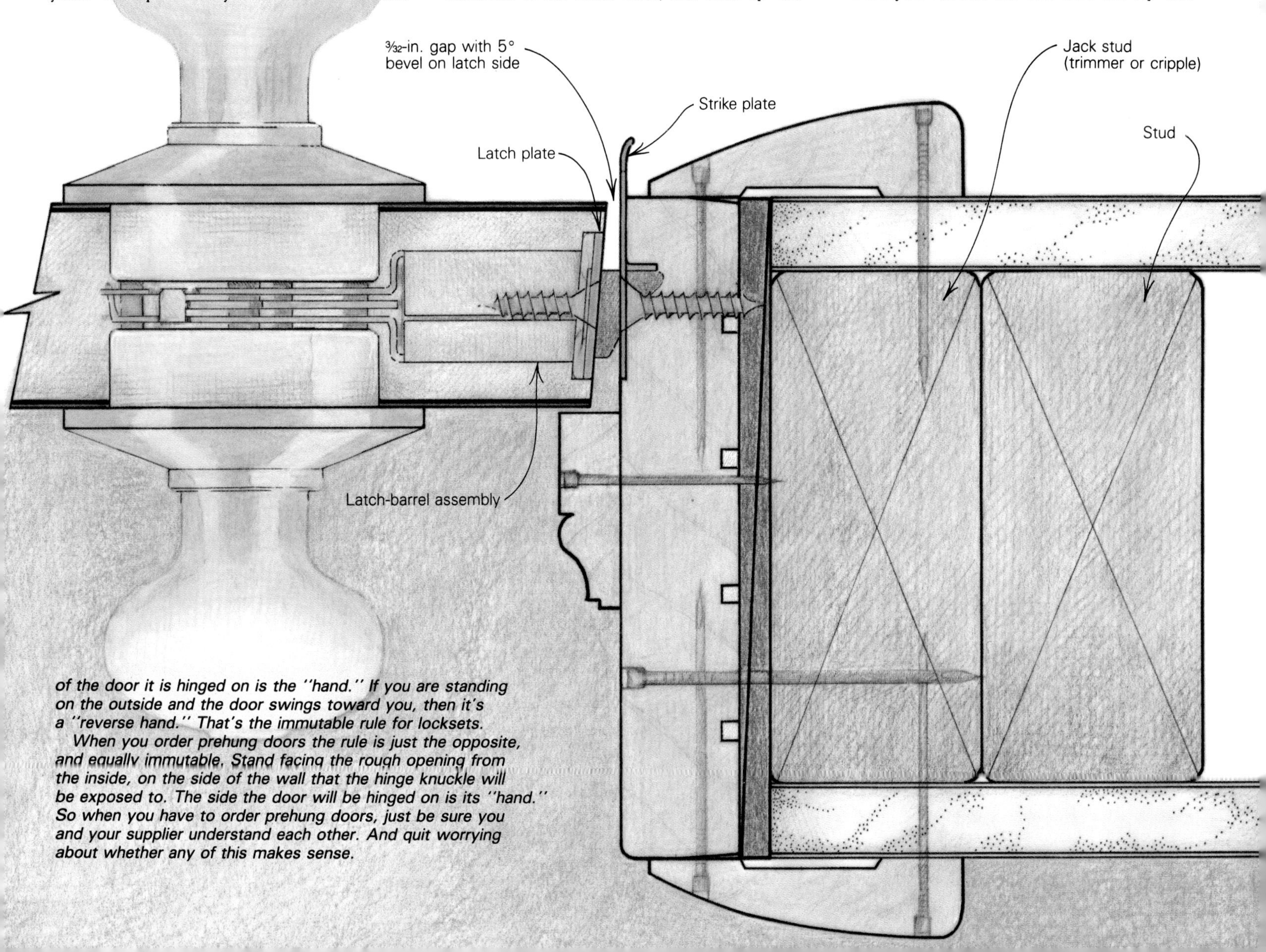

of the door it is hinged on is the "hand." If you are standing on the outside and the door swings toward you, then it's a "reverse hand." That's the immutable rule for locksets.

When you order prehung doors the rule is just the opposite, and equally immutable. Stand facing the rough opening from the inside, on the side of the wall that the hinge knuckle will be exposed to. The side the door will be hinged on is its "hand." So when you have to order prehung doors, just be sure you and your supplier understand each other. And quit worrying about whether any of this makes sense.

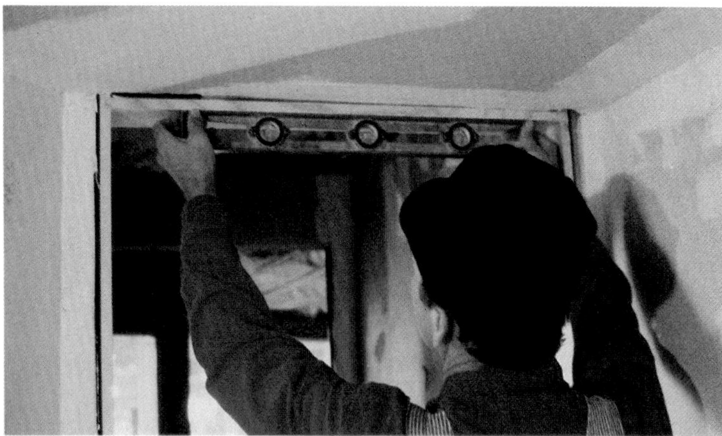

Setting the jamb. **After standing the assembled jamb in the rough opening, try the head for level, as shown. Then adjust it if necessary by shimming up the side jambs.**

Here the side jambs need to be trimmed to rest on oak flooring. A piece of scrap equal to flooring thickness is placed beneath each side jamb as level adjustments are made with a scribe. When the scribe is sawn off and the jamb positioned in the opening, the head will be level.

To get the side jambs parallel, hang a plumb bob from the center of the head jamb and shim the side jambs right or left to get the bob centered over the spreader at the bottom.

Finally, to straighten the sides, use a long straightedge against the jamb and drive in wedges (shim shingles) to bring the jamb into alignment. Be certain to position the wedges at hinge locations on the hinge side.

From *Fine Homebuilding* magazine (April 1985) 26:26-32

two 8d finish nails. Nail right through the shim wedges. Check the other side for alignment and nail it. Now go to the bottom, and nail both sides. At this point the chances are that the side jambs won't be straight top to bottom, but will be bowed slightly in or out.

To correct this condition, lay a straightedge (one just shorter than the height of the opening) against one side of the jamb. Lean against the straightedge with your body, and wedge a pair of shim shingles between the jamb and jack stud to bring the jamb out to meet the straightedge (photo facing page, right). Hold the jamb against the wedges and remove the straightedge; give the wedges just a hair more push, and then nail the jamb right through them with two 8d finish nails. The additional push of the wedge moves the jamb slightly out of alignment, but the nailing will compress the small wood-to-wood air spaces, and bring the jamb back into line. Return the straightedge to the jamb and adjust where necessary. Once both jambs have been straightened, they will be parallel.

On the hinge side, be sure to put the wedges near the hinge locations. This will hold the jamb firmly when it is mortised for the hinges and will transfer the weight of the door directly to the wall framing.

After the jamb is nailed up, you can use a handsaw to cut off the protruding wedges, or score them on both sides with a utility knife and snap them off. Be sure they are behind the wall surface so they won't be in the way of the casing when it's applied.

The next step is applying the door stop. Most interior doors are 1⅜ in. thick, so use a combination square and a pencil as a marking gauge to scribe a line 1⅜ in. in from the latch side of the jamb. On the hinge side, this line should be 1⁷⁄₁₆ in. in from the edge of the jamb. The additional ¹⁄₁₆ in. is necessary to prevent the door from binding or stopping at the end of its swing.

The corners of the door stop can either be mitered or coped; coping is usually better. As a rule, I cut the stop and tack it in place, but I don't nail it up until after the door is hung because it may be necessary to make adjustments to get the door to close properly. Also, I have to remove the stop on the hinge side to accept the hinge-mortising templates. The casing can be applied now or after the door is hung.

Fitting the door—Gone are the days when a heavy door jack was used to hold the door on edge while a carpenter shaved it to size with a jointer plane. While it's fun to swish off 7-ft. long curls of pungent pine with a hand plane, using a power plane is faster, and it takes much less effort. Instead of a door jack, I use a door bench with enough room on it to hold the tools I use—a power plane, a router for hinge mortising, a skillsaw and cutoff guide, a belt sander, a drill, a jack plane, a Stanley No. 271 router plane, which I use to mortise for latch plates and strikes, and my Yankee screwdriver.

I've known carpenters who could place a door in an opening, mark all the margins and mark the hinge and lock locations in one trip; then they could cut the door to size, mortise for hinges and lockset and return it for hanging without further adjustments. I'm seldom that lucky, and the method I use allows for a trial fitting and a final trimming.

The traditional margins for a door are ¹⁄₁₆ in. of space between jamb and door at the head, ¹⁄₁₆ in. on the hinge side and ³⁄₃₂ in. on the latch side. Old carpenters used to call it nickel and dime—a nickel on each side and a dime on top. They used the coins as feeler gauges; a nickel is a big ¹⁄₁₆ in. and a dime is a thin ¹⁄₁₆ in. Since ¹⁄₁₆ in. is the smallest graduation on a carpenter's rule, thirty-seconds are referred to as big and little sixteenths in carpenter-speak. The old rules make a door a little too tight in my opinion, so when I fit doors, I use a nickel on top, ¹⁄₁₆ in. on the hinge side and ³⁄₃₂ in. on the latch side. The space at the bottom is from ⅛ in. to ½ in., or large enough to clear any high spots on an out-of-level floor.

Doors, like other things made of wood, shrink and swell with changes in season. The part of the country where I live has wide swings in humidity and heat from summer to winter, and the time of year has to be taken into account. A door fitted tightly in winter will strike the jamb when closed in summer, and a perfectly fitted summer door will be drafty in winter. In the summer, fit close; in the winter fit loose. It's a common mistake to recut the margin on a door during its first summer, only to have it too small when winter returns. If a door must be planed down, reduce it only enough to clear. Also, don't plane the latch side—if you do, the lockset will need to be shifted and some of them can't be moved without exposing the holes. Plane the hinge side, then deepen the mortises and reset the hinges.

Fitting a door must proceed step by step. Lay the door across the top of the bench or on a pair of sawhorses, and cut it just small enough to fit snugly in the opening. If you have to remove more than a ⅜-in. wide strip, use your skillsaw with a sharp planer blade. If you have to saw off the bottom of a veneered door, first score the line of cut with a sharp utility knife. This will keep the veneer from tearing and chipping out. But if you're taking the edges of the door down by ¼ in. or less, it's easier to use a power plane (the direct-drive type is designed for edge planing). After trimming the door to fit the opening, place it in the jamb and force it up against the head by wedging from the floor and over against the hinge side by wedging from the latch side. Scribe a line across the head ¹⁄₁₆ in. below the jamb and another line the required distance up from the floor.

On the latch side, measure over from the jamb ³⁄₃₂ in. or a fat ¹⁄₁₆ in., and mark it on the door. And on the hinge side make a mark ¹⁄₁₆ in. over from the jamb. Remove the door from the jamb, set your combination square by the mark on the latch side and mark a line down the entire length of the door. Flip it over and do the same on the hinge side. Secure the door on edge and plane to the line on the hinge side and on top and bottom.

Now you can trim and bevel the latch side of the door. Flip the door over so the latch side is up and the scribed line toward you. This edge must be planed to the line and beveled. As a door swings on its hinges, the leading edge will strike the jamb if it's not beveled. The degree of bevel can vary, but I find that 5° is good in most cases. So when setting the fence on my power plane, I adjust it by eye about 5° out of square. I set the plane's depth of cut to about ¹⁄₃₂ in. and make successive passes from one end to the other until I shave off the top of the line.

Return the door to the jamb, wedge it in place and check the margins on all four sides. If any side needs a slight trim, plane it down and check again. Once the fit is correct, I take the arris off all of the edges of the door with my smooth plane. Sharp 90° edges on a door can be dangerous.

Mortising for hinges— To cut hinge mortises in the door and the jamb, I use a router and a pair of site-made templates connected by a bar. This setup eliminates the need to lay out the hinge locations on each door and every jamb, and it makes for quick, accurate work. This can be especially important if you have a houseful of doors to hang. But if you're hanging only one or two doors, you might want to cut the mortises with the chisel in the traditional way that's described in the sidebar on p. 21. Manufactured hinge-mortising templates are available, but I prefer to make my own because it's less expensive, offers me greater flexibility and reduces the amount of permanent gear and dinky parts I have to keep track of and take from job to job. All that's required using my method is a sharp butt chisel, a router, a ½-in. straight-face bit (preferably carbide tipped) and a ⅝-in. O.D. guide bushing, sometimes called a template guide, for the router.

I begin by laying out a single door and its jamb for hinges. Place a nickel on top of the door and wedge it in place (you can do this at the same time you check the margins for accu-

The edge on the latch side of a door must be beveled about 5° if it is to close without striking the jamb. This can be done with a hand plane or by using a power plane, as shown at right. The fence is set about 5° off the perpendicular, and the depth of cut is set for about ¹⁄₃₂ in. Successive passes are made until the scribed line on the open side is reached.

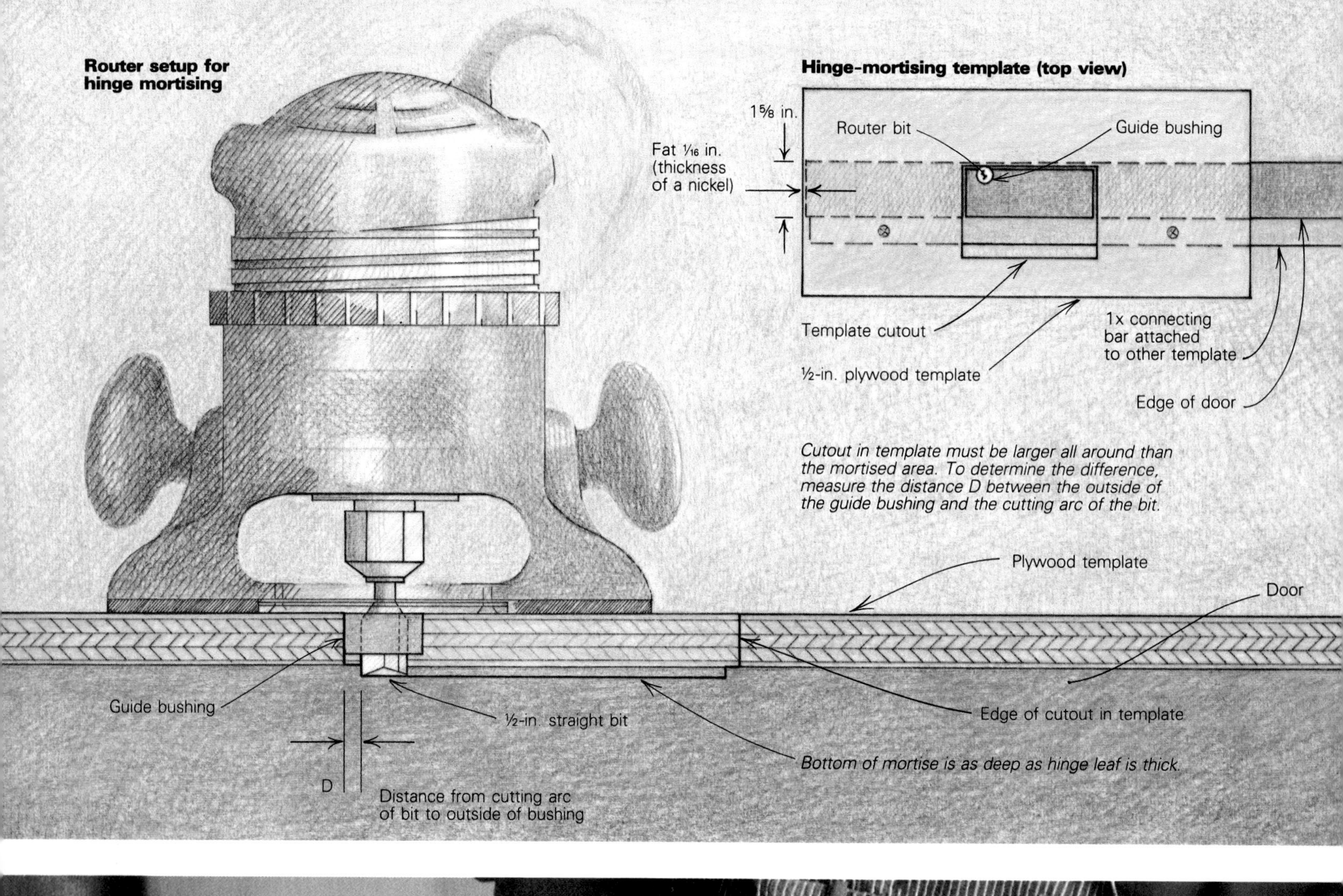
Router setup for hinge mortising
Plywood template
Door
Guide bushing
½-in. straight bit
Edge of cutout in template
Bottom of mortise is as deep as hinge leaf is thick.
D
Distance from cutting arc of bit to outside of bushing
Hinge-mortising template (top view)
1⅝ in.
Router bit
Guide bushing
Fat 1⁄16 in. (thickness of a nickel)
Template cutout
½-in. plywood template
1x connecting bar attached to other template
Edge of door
Cutout in template must be larger all around than the mortised area. To determine the difference, measure the distance D between the outside of the guide bushing and the cutting arc of the bit.

BIRCH

racy); then use a straightedge to strike a line across door and jamb simultaneously. Then mark the side where the hinge will be. That seems simple enough, but many a hinge has been confidently mortised into the wrong side of a door. Mark the location for the top of the top hinge and the bottom of the bottom hinge, and the center of the center hinge if your door has one. The top of the top hinge should be 5 in. from the top of the door, and the bottom of the bottom hinge 5 in. from the bottom of the door.

With the door on edge, hinge side up, draw the lines you marked on the face across the full edge of the door. Position one of your hinges on the correct side of the line so that the center of the barrel or knuckle is about ½ in. from the edge of the door. This approximation is for standard 4x4 butts; distance will vary slightly depending on door thickness, hinge size and clearance needed to keep the door from contacting the casing when opened a full 180°. Positioned properly, the back edge of the hinge will lie about 3⁄16 in. from the back edge (the closed side) of the door. Holding the hinge firmly on the layout line, trace around the back edge and opposite side with a sharp pencil, or score lightly with your penknife.

Next, cut a pair of rectangles about 14 in. long and 6 in. wide from a scrap of ½-in. plywood. These will be your router templates, and into them you'll jigsaw rectangular cutouts that will guide the bushing on the router to produce a precise mortise in the edge of the door (and jamb) that will hold the hinge leaf.

To get the mortise the right size, the template opening will have to be larger than the hinge outline by a specified amount. As shown in the drawing, facing page, this amount equals the distance from the outside of the guide bushing to the cutting arc of the bit. So to lay out the template, trace around the hinge leaf, and enlarge this outline on all three sides in the correct amount. Lay out the templates for both hinges so that the top edge of the finish cut for each is 5 in. from the outer edge. Leave enough room on the open side of the template to maneuver the bit up to the edge of the door. Now saw just shy of the line, and then pare right down to it with a sharp chisel.

Mount one template over a trial hinge layout on the edge of a length of 2x. Secure the template with a couple of 4d finishing nails; then make a trial cut with the router.

Now remove the template (leaving the nails in their holes) and square up the radiused corners of the test mortise with your chisel. Place the hinge in the mortise to check for a snug fit. If it's too tight and the hinge must be forced in with a hammer, enlarge the cutout in the template slightly with a bastard file and check the fit again. If the hinge is too loose it may not stay square, which will cause it to be out of alignment with its mate on the jamb. To remedy this condition, you can add veneers of masking tape to close up the space.

The depth of the mortise should be exactly the same as the thickness of the hinge leaf. The old-timers used to say, "It looks like it growed there." But there are some instances, which I'll discuss further on, when the hinge depth needs to be varied. To set your router bit the proper depth, turn the router upside down, and gauge the depth of cut by holding the hinge leaf on top of a small scrap of ½-in. plywood. Make another trial cut to make sure that the hinge will fit flush with the wood around it.

With the two templates complete and the router bit set to the proper depth, position the top-hinge template over the hinge layout on the edge of the door. Because the template for the top hinge will fit right against the head when you mortise the jamb, it needs to be flipped end for end and then offset when you mortise the door to produce the margin between door and head. As shown in the photo, facing page, I hold a nickel on the top edge of the door and bring the top edge of the template out flush with it to get the correct spacing. But make sure that the back of the cutout is positioned parallel to the layout line at the rear of the mortise, and that the spacing between templates is correct. Now nail the second template over its layout at the bottom of the door.

Now get a straight 1x ripping that's about as long as the door, and attach it to the underside of both templates (with drywall screws and glue), making certain that it fits closely against the edge of the door. This bar serves to hold the two templates the proper distance apart throughout the whole job, and it acts as a fence, ensuring that each mortise is the correct width by holding the back edge of the cutout at a constant distance from the edge of the door.

After mortising the door, take the template to the hinge side of the jamb, index the top of the top-hinge template hard against the head jamb and tack it in place, top and bottom. Then rout the mortises (photo below right).

Installing the hinges—When the routing is done and you've squared up the inside corners with your chisel, place the hinges in the door mortises (make sure barrels are oriented pin up) and mark the centers for the screw pilot holes. For this I use a nail set because it's blunt and crushes its way into the wood fibers. A sharp punch or an awl can be pulled off center by coarse grain. I don't make the pilot hole exactly in the center of the screw hole, but just a hair off center, in toward the back (closed end) of the mortise. Doing this will pull the hinge leaf snugly against the back wall of the mortise when the screws are driven home.

Phillips-head screws are better for hinges than slotted screws. The Phillips screws can take a lot of torque without letting the bit slip out of the slot and burr the head of the screw and scar the hinge leaf. Door men use a Yankee screwdriver because it saves lots of time, keeps your hand from aching and your forearm from cramping. To ease the friction while driving, drag the screw threads though a bar of wax made by melting and mixing paraffin and beeswax. The spiral on a Yankee screwdriver requires lubricating from time to time, but don't use oil, which will splash on the work, get on your hands and wind up everywhere. Paraffin is good for this and it won't make a mess.

Hanging the door—Now that the door is trimmed to size and the hinges mounted, it's time to hang it on the jamb. One method is to split the hinges and with one leaf on the door and its mate on the jamb, lift the door, slide the knuckles together and insert the pins. This works fine if the door is light, if there are only two hinges and if their alignment is perfect. If there are three hinges and their alignment is just less than perfect, hang the door on the top and bottom hinges and then mount the center hinge in place with the door open.

Close the door and check the margins—they should be perfect all around, 3⁄32 in. on the latch side, 1⁄16 in. on the top and 1⁄16 in. on the hinge side. There are, unfortunately, occasions when this observation causes your heart to sink. All that careful work and look at it. The top's not parallel to the head, and on the latch side it's closer on the top than at the bottom, and on the hinge side it's closer on the bottom than the top. Despite your despair, this condition is not unusual, and it can be corrected.

Heavy doors can sag once they're hung because they take up the slight play between the knuckles and pins. To compensate for the sag, the hinges must be adjusted. There are several ways to do this. You can pull the top in closer by making the mortise in the door just a little deeper than the leaf thickness. Or you can push the bottom out by shimming behind the bottom hinge with a piece of compressed cardboard. (The box the hinges came in is about right.) Remove the bottom leaf from the jamb and insert a leaf-size piece of cardboard behind it. Repositioning the hinge over the cardboard will push the bottom of the door over and move the top edge up. Usually this is enough to take the sag out, but if it isn't, you can pull the top over by

Mortising the door. **Law tacks site-made hinge-mortising templates to the edge of a door (facing page). He uses a nickel as a feeler gauge to position the template a fat 1⁄16 in. beyond the top of the door.**

Mortising the jamb. **Because the template will fit hard against the head when the jambs are mortised (right), the offset on the door will produce the proper gap between door and head jamb. The whole jig has to be flipped end for end when routing the jamb, so the layouts have to be symmetrical. Before routing the jamb, make sure shim shingles are directly behind the mortise to transfer the weight of the door to the studs.**

mortising a little deeper or by "throwing" the hinge to move its pivot point.

Looking down at a section drawing taken through the door, hinge and jamb, the center line of the margin passes through the center of the hinge pin. "Throwing" this pin center left or right has the effect of moving the door left or right. To do this use cardboard shims as before, but this time cut them into narrow strips about 1/4 in. wide. By placing one of these strips between the jamb and leaf at the rear or closed side of the mortise, you will throw the hinge to the right. Putting a shim strip on the open side of the mortise will throw the hinge to the left.

Quicker by the dozen—When I'm hanging a house full of doors I follow this routine: First I sort the doors and jambs for size or type. Then I cut the jamb heads to size, nail them to the sides, and take each assembed jamb to its rough opening where I lean it against the wall. Generally I make the jamb width the nominal door size, such as 24 in., 30 in., or 36 in. But you should be sure of the door size, because some of them are already cut down. Using the technique described above for setting the jambs, I start at one end and go to one jamb after another until they're all set. Next, I cut the stops and tack them in place. Then I start at the beginning again, and using the plywood router template, I mortise all the jambs for hinges.

At this point I have not applied any of the casings because they get in the way of the mortising template. Also, if any adjustments to the jamb are needed after the door is hung, I can still get to the wedges.

With the door on the bench and latch side up, I plane it to size and cut the bevel. Since I know the size of the opening and know the sides are straight, I just plane the door to 5/32 in. less than the width between the jambs on the bench. Then I flip it over and rout out the mortises. Then I install the hinges, take the jamb leaf to the jamb and mount it, carry the door to the jamb, hang it on the top hinge, open the door and drive the screws on the bottom hinge. I close the door and check the margin. No heartaches. The casing can now be applied. The next step is to go through and drill out and mortise for the locksets, and then installing them. When the doors are hung and locksets installed (sidebar at left), the stops can be adjusted to fit those doors that are slighly warped.

By moving from door to door with each operation, you can save a lot of time because you have each tool you need for each repeated operation. Using this production-line method, I can approach the speed of installing prehung doors, but still my way costs more. Is it worth it? I think that it is, because I can achieve better results than what comes from the mill. Many of the assembly-line workers in mills do shoddy work, and quality control is down. I've seen every possible mistake in doors and jambs that come as units from the mills. Often it's easier to do the job right yourself than to fix what another person has fouled up. □

Consulting editor Tom Law is a carpenter in Davidsonville, Md.

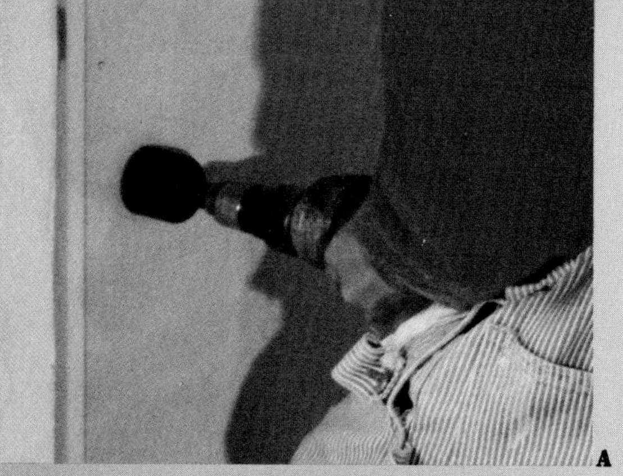
A

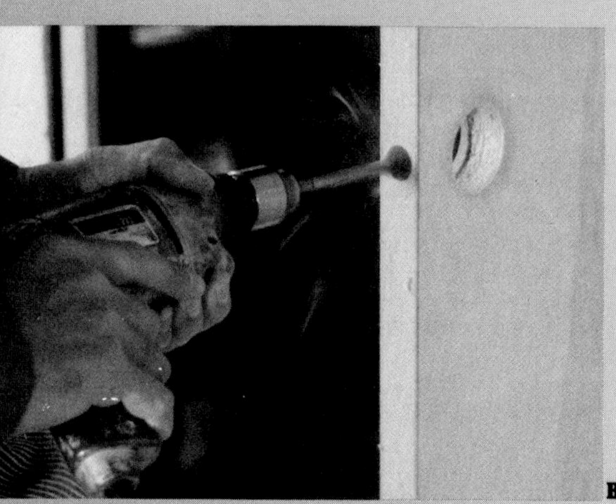
B

C

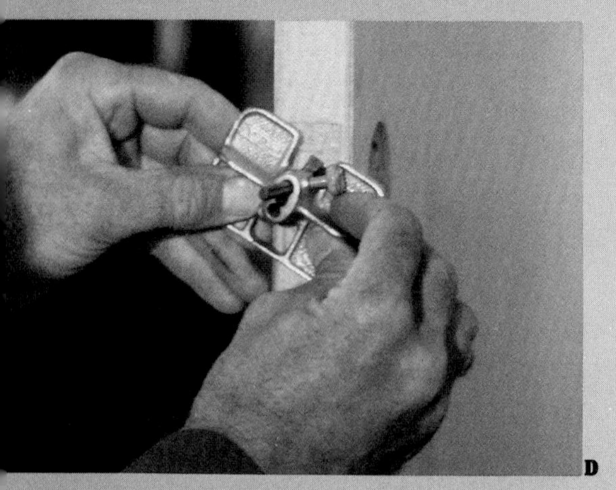
D

E

Installing a lockset

Most interior locksets consist of three parts—a pair of knobs, a latch assembly with integral latch plate and a strike plate, which is mounted on the jamb. To install these three items you bore three separate holes. They must all relate to one another in a precise and prescribed way. Some locksets come with templates that help you locate the three bore centers; others come only with instructions that tell you where to position the holes relative to the edges of door and jamb.

You begin by laying out the bore centers for the knobs and the latch-barrel assembly. Then you lay out the center for the strike-plate bore on the jamb. To locate the center for the knobs, measure up from the bottom of the door 36 in. to 39 in. (depending on your preference) and mark a light line with your combination square across the edge and both faces of the door. Then, using the square as a depth gauge, mark a vertical line 2 3/8 in. or 2 3/4 in. (depending on the lockset you have) back from the edge of the door. Next find the center for the latch-barrel bore by dividing the line on the door's edge precisely in two. Finally, mark the center for the strike-plate bore by finding the point on the jamb exactly opposite the center on the door for the latch bore.

After punching all these centers, the next step is to cut the hole for the knob assembly (A). I do this with a 2 1/8-in. hole saw (standard for most locksets), though you can use an expansion bit in a hand brace. Saw about halfway through on one side and then finish the hole by cutting from the other side. Next, chuck up a 7/8-in. dia. spade bit in your drill and bore the hole for the latch barrel (B). Be sure to hold the bit level so that the bore breaks through into the center of the knob hole. Next, using the same bit, bore out a 5/8-in. deep hole in the jamb for the strike plate.

Now you can lay out and cut the shallow mortises for the latch plate and the strike plate. For this I use a 1-in. chisel and a little Stanley No. 271 router plane. Begin by inserting the latch barrel into its hole and squaring up the plate with the edge of the door. Trace around it carefully with a sharp pencil or a striking knife. Remove the latch, and cut the wood at the top and bottom of the layout by giving the chisel a smart tap with your hammer (C).

Then do the same along the sides, though you have to be careful here not to drive the chisel too deeply and risk splitting out the wood along the edge of the door. Set the cutter on the router plane to the thickness of the latch plate and remove the wood inside the mortise layout (D). With a router plane you can quickly excavate the mortise to a uniform depth all around. The router plane also ensures that the bottom is flat. Repeat this procedure on the jamb for the strike plate, and you're ready to install the latch (E). —*T. L.*

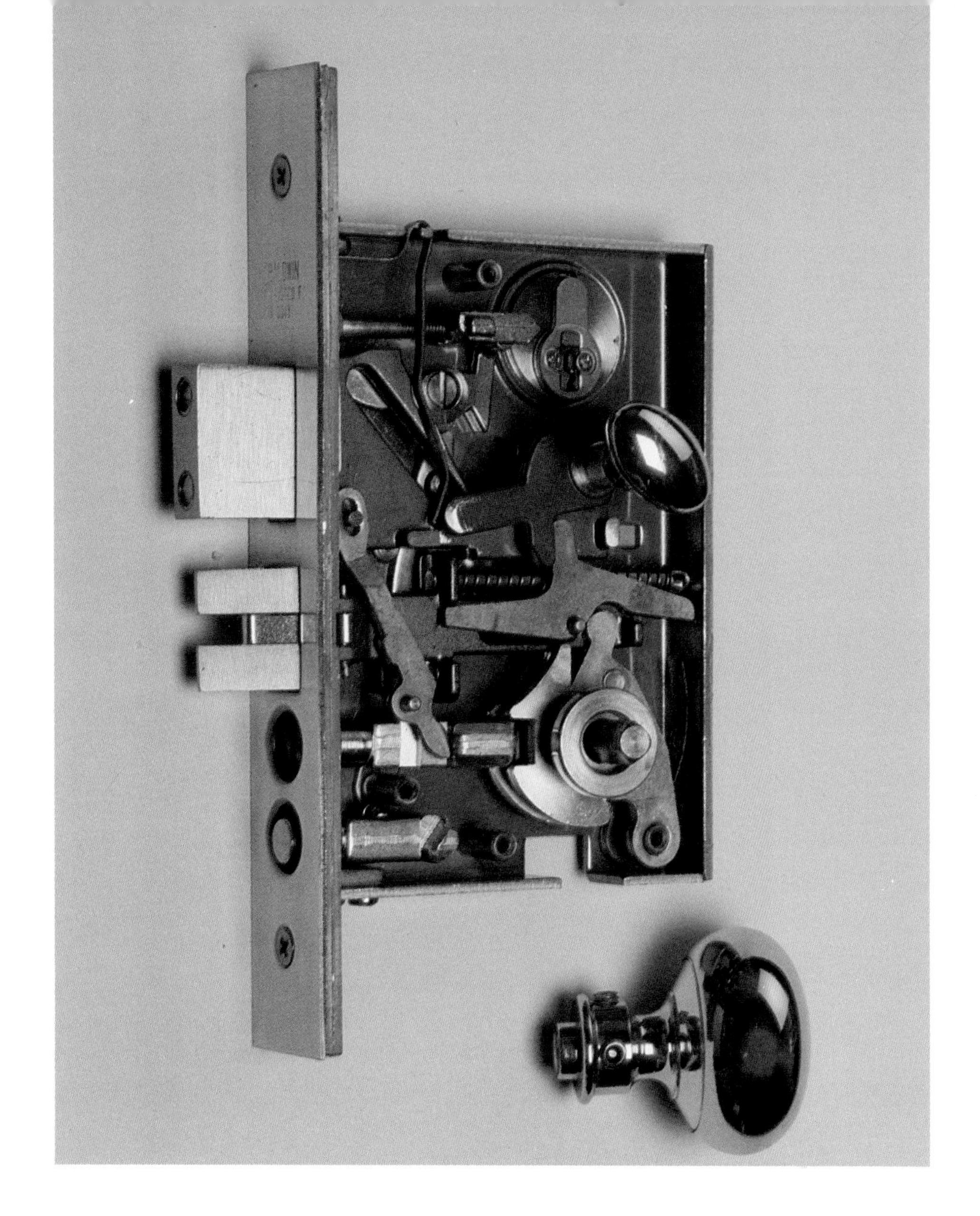

Door Hardware

Getting a handle on locksets, latches and dead bolts

by Kevin Ireton

Walter Schlage, a German immigrant working in San Francisco, obtained his first patent in 1909. The idea came to him late one night, as he unlocked his front door and reached in to turn on the lights. The patent was for an electrified door lock that automatically turned on the lights inside a house when the front door was unlocked.

You can't relate the history of the lockset industry without talking about Schlage, Dexter and Yale, men whose companies are still in business. But today, more than 50 other companies now manufacture locksets and dead bolts. From 18th-century reproductions to ergonomically designed lever handles, more door hardware is available today than ever before.

Most door hardware performs at least two basic functions—latching and locking—and many aspire to a third function—looking good. The term lockset refers collectively to the complete latch bolt assembly, trim and handles (knobs or levers). A latch bolt is a spring-loaded mechanism that holds a door closed and may or may not have a lock incorporated into it. A dead bolt, on the other hand, is not spring-loaded and can only be operated with a key or a thumbturn.

Rim locksets and mortise locksets—Originally just wooden bolts that slid across the opening between door and frame, rim locksets were the first broad category of locksets to evolve. They developed into the surface-mounted wrought iron and brass cases common in colonial America.

Eventually, locksmiths realized that doors would look better if the locksets were out of sight. So they mortised the rim lockset into the edge of a door to create the mortise lockset, which prevailed throughout the 19th century. Mortise locksets are still widely used, especially in commercial installations. In residential construction, they're used for exterior doors more ofcn than for interior, though models for both uses are available.

Some people consider mortise locksets to be the best type of door hardware available (photo above). Manufacturers can build more

From *Fine Homebuilding* magazine (August 1988) 48:61-65

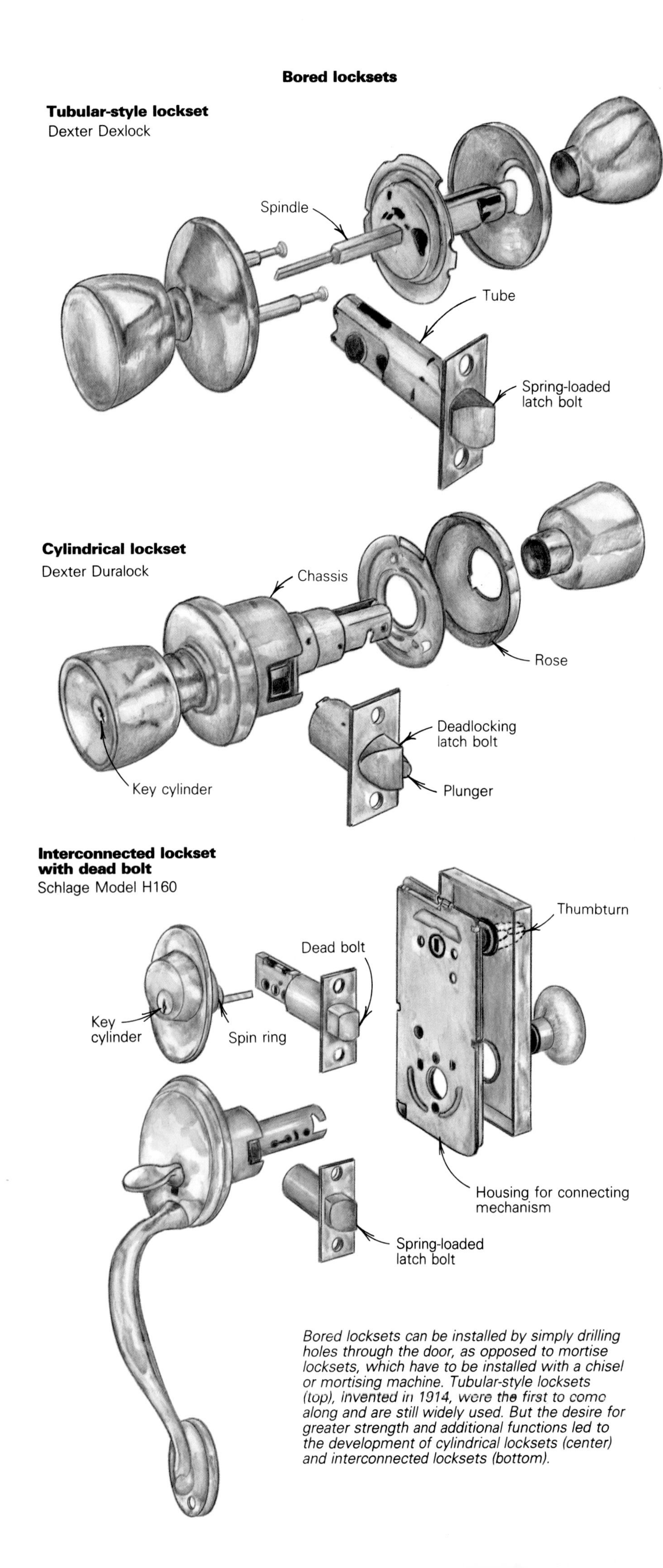

Bored locksets can be installed by simply drilling holes through the door, as opposed to mortise locksets, which have to be installed with a chisel or mortising machine. Tubular-style locksets (top), invented in 1914, were the first to come along and are still widely used. But the desire for greater strength and additional functions led to the development of cylindrical locksets (center) and interconnected locksets (bottom).

strength and longer wear into the larger cases than they can into the bored locksets that I'll discuss later. Also, because the latch and dead bolt are housed together, they can be interconnected to offer a variety of functions. Turning the inside knob on a mortise lockset, for instance, retracts both the latch and the dead bolt—no fumbling for the dead-bolt thumbturn if you're in a hurry to get out. The push buttons, called stop-works buttons, are located below the latch on most mortise locksets and determine whether the latch bolt can be retracted by the outside knob. This type of lockset is also easily adapted for use in extra-thick doors. All that's needed is a longer spindle for the knob.

Aside from the fact that they cost more than other types of door hardware, mortise locksets carry the disadvantage of being tough to install. Cutting a mortise 6-in. long, 4-in. deep and nearly 1-in. wide in a door that's only 1¾-in. thick is tricky and time-consuming. The standard method involves drilling a series of closely spaced holes in the edge of the door and chiseling to clean out between them.

To ease the job, Porter-Cable Corporation (P. O. Box 2468, Jackson, Tenn. 38302-2468) makes a tool called a lock mortiser (model 513). It clamps to the edge of the door and has an integral motor that holds a long mortising bit, much like a router does. Turning a crank on the side of the unit moves the bit up and down along the door and advances the depth automatically. Once it's set up, the lock mortiser works fast, but at a suggested list price of $975, you'll have to mortise a lot of doors to pay it off.

Bored locksets—In 1914, a Michigan hardware manufacturer named Lucien Dexter realized that installing mortise locksets on every door in a house amounted to overkill—nobody needed that much hardware just to keep the pantry door shut, so Dexter invented the tubular-style lockset.

The tubular-style lockset is basically a spring-loaded latch bolt, housed in a tube, that mounts in the edge of a door (top drawing, left). It's operated by a spindle that runs through the end of the tube. A smaller, simpler mechanism than the mortise lockset, the tubular-style lockset is also much easier to install. All you have to do is drill two intersecting holes in the door, one through the face—the crossbore—and the other through the edge—the edgebore (for more on installing bored locks, see p. 28).

The drawback to the tubular-style lockset is that most of the working parts are contained in the narrow tube, so there isn't room to incorporate many functions or a very secure lock. Walter Schlage overcame this problem with his next invention in about 1924, when he developed the cylindrical lockset. Like the tubular-style lockset, the cylindrical lockset incorporates a spring-loaded latch bolt mounted in the edge of the door. But instead of just a spindle passing through the tube, a bigger crossbore hole is drilled (usually 2⅛-in. dia.), and a large chassis is

Drawings: Michael Mandarano

inserted through the door. The cylindrical lockset and the tubular-style lockset are classified together as bored locksets because both are installed by boring holes.

Before the invention of the cylindrical lockset, latching and locking were performed by separate mechanisms. But the large chassis on his new invention gave Schlage enough room to install the key cylinder right in the knob, so that locking and latching could be combined in one mechanism (for more on key cylinders, see sidebar this page). Schlage also developed the button-lock for the inside doorknob, an idea he got from push-button light switches in use at the time.

Among later refinements to the cylindrical lockset was the deadlocking latch bolt. The familiar trick of retracting a latch bolt by slipping a credit card between the door and jamb is known in the trade as "loiding a lock," *loid* being short for celluloid. The deadlocking latch bolt thwarts this practice. With the door closed, a spring-loaded plunger, located alongside the latch bolt, is held in by the strike plate, and the latch bolt can't be retracted. This requires a reasonably conscientious installation. Otherwise, the plunger can slip into the hole for the latch bolt, which defeats the purpose.

Dead bolts—"Don't rely on a lockset for security." I've heard that from a number of people in the industry. Projecting out from the door like it does, a key cylinder in a doorknob is vulnerable to being smashed or sheared off. If you want security, you have to have a dead bolt. Some insurance companies even offer you a 2% or 3% discount on homeowner's insurance if you have dead bolts on all your exterior doors. Still, I was reminded by a locksmith that, "One-hundred percent security is not available on this planet. Anything that's manmade can be defeated. By using dead bolts, you're simply trying to make access more difficult."

Dead bolts should have a full 1-in. throw, which means the bolt will extend 1 in. into the door jamb. At one time, shorter throws were common, and some companies may still make them. But if the dead-bolt throw is less than 1 in., it's easy for an intruder to prize the door jamb away from the door and free the bolt.

Many companies run steel inserts through the center of their dead bolts so that if someone tries to hacksaw them, the inserts will roll under the blade instead of allowing it purchase to cut. It's a great idea in theory, but no one I asked had ever heard of a burglar hacksawing a dead bolt; there are too many easier ways to break into a house. Peace of mind, I learned, is a major factor in selling home security.

Most dead bolts include a key cylinder outside and a thumbturn inside the door. But dead bolts are also available with a double cylinder, which means you need a key to unlock it from either side of the door. Double-cylinder dead bolts are used in doors that have windows so an intruder can't break the window, reach in and unlock the door. In emergencies, though, they can pose a safety hazard, and their use is restricted by code in some areas.

The weak spot in a dead-bolt installation is the strike—the metal plate mounted in the door jamb that the dead bolt goes into. It should be mounted with 2-in. or longer screws that penetrate the trimmer stud behind the jamb. Some companies supply long screws with their dead bolts, but some don't, and often you have to angle the screws to sink them solidly into the framing. But now a few manufacturers have begun to offer dead-bolt strike plates with holes that are offset toward the center of the door jamb.

Many dead bolts come with a metal or plastic housing, called a strike box or dust box, that installs behind the strike and keeps the dead bolt contained inside the jamb. Installing it requires extra work with a chisel, and I'll admit to throwing away more strike boxes than I've installed. I thought their only purpose was to make the hole in the jamb look neat. But it turns out that strike boxes serve a couple of important purposes. For one thing, they keep debris from getting into the strike hole and jamming the bolt. But

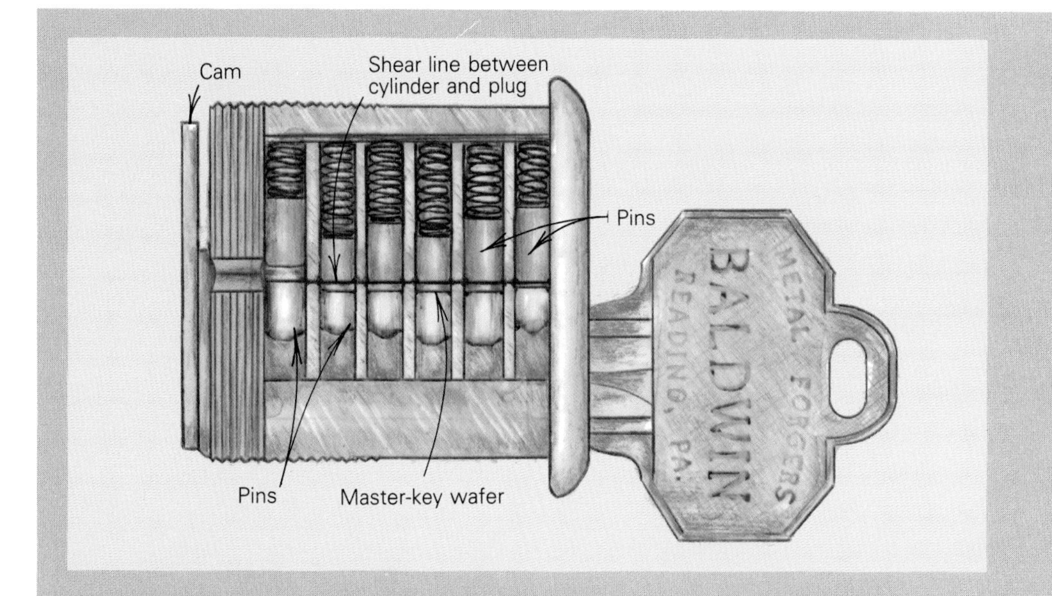

Pin-tumbler cylinders

The oldest known lock, found at the ruins of the palace of Khorsabad in present-day Iraq, is over 4,000 years old and operates on the same principle as most door locks today—the pin-tumbler. This lock and others like it were operated by a wooden key shaped something like a large toothbrush, but with wooden pins in one end instead of bristles. When the key was inserted and lifted up, the pins on the key aligned a set of pins inside the lock that allowed a wooden bolt to be retracted. This type of lock was common in Egypt and is called an Egyptian lock. Subsequent civilizations developed new locks based on different principles, and pin-tumblers weren't used again until 1848 when Linus Yale "reinvented" them.

Yale developed and patented a simple key-operated lock based on the pin-tumbler principle. His son later refined the lock by encasing it in a small metal cylinder with a rotating internal plug. Although plenty of locks have been invented since—combination locks, push-button locks, credit-card locks and even electronic locks—the pin-tumbler cylinder is still the heart of most door locks used today.

***How it works*—A series of holes, usually five or six for a door lock, are drilled through the top of the cylinder into the plug. At least two pins of different lengths, plus one spring, are inserted into each hole. Some cylinders have brass pins, but the best cylinders use nickel silver, which is a harder metal than brass and assures that the key will wear out before the pins. By varying the combinations of pin lengths, a manufacturer can key any given cylinder literally millions of different ways.**

The line between the plug and the cylinder is called the shear line. Any misalignment of the pins across the shear line prevents the plug from rotating in the cylinder, and hence, the lock from unlocking. But when the right key is inserted in the keyway, the pins slip into the gullets of the key's serrated edge, which raise them to just the right height, aligning their ends along the shear line (drawing above). The plug is then free to turn in the cylinder, and when it does, a cam attached to the back of the plug throws the bolt.

A master key is one that will work two or more cylinders that are keyed differently. Cylinders accommodate a master key by means of master-key wafers inserted with the regular pins. These wafers create additional points where joints between the pins align at the shear line and the plug is free to move.

Most key cylinders are assembled by women because the pins are very tiny, and women, with their smaller hands, can assemble cylinders much faster than men can. —*K. I.*

their main function is to serve as a depth gauge. If the dead bolt is less than fully thrown, only ¾ in. for instance, it will easily spring back into the door when you tap it—not a very secure situation. Installing the strike box ensures that the dead bolt can be fully thrown.

The large beveled housing surrounding the outside of a dead-bolt cylinder is called a spin ring. It protects the key cylinder from being removed with a pipe wrench or locking pliers. When buying a dead bolt, be sure to get one that has a solid metal insert for the spin ring. Hollow rings can be crushed easily.

Weiser Lock (5555 McFadden Ave., Huntington Beach, Calif. 92649), makes a combination lockset and dead bolt called the Weiserbolt. It works like a standard lockset, until you turn the key in the lock or twist the thumbturn. Then the latch bolt extends a full 1 in. into the jamb. But while the Weiserbolt is more secure than a standard lockset, it isn't as secure as a separate dead bolt because the key cylinder in the projecting knob is still vulnerable. It does, however, provide additional security without requiring an extra hole in your door.

Interconnected locksets—In order to combine one of the functions of a mortise lockset with the easy installation of a bored lockset, the Schlage Lock Co. introduced the interconnected lockset in 1967. It's an entrance lockset that links the latch and dead bolt so that both will retract by simply turning the inside handle. The mechanism connecting the dead bolt and latch bolt is contained in a thin rectangular housing that mounts on the inside face of the door (bottom drawing, p. 30). Interconnected locksets cost more than separate locksets with dead bolts, but they are more convenient to use.

Interior locksets—Locksets for interior doors don't usually have a keyed lock, but may instead have a simple button lock. These are called privacy locksets and are used for bedrooms and bathrooms. Privacy locksets have an emergency release so that a locked door can be opened from the outside either with a special tool that comes with the lockset or with a small screwdriver.

A passage lockset is one with no locking function at all; turning the knobs or lever handles on either side of the door will always retract the latch. Some manufacturers offer a special closet lockset that has a knob or lever handle on the outside, but a simple thumbturn on the inside. These are usually available only in more expensive lines, because if you're paying $30 or $40 for a solid brass doorknob, you may not want waste it inside a closet. A dummy knob is a handle with no lockset and is often used on the stationary half of a pair of French doors.

The photo above shows the stages a lever handle goes through in the hot-forging process. Once cut to length, pieces of bar stock, called billets, are heated to 1,400° F. The red-hot billets are placed in a die, slammed by a forging press, and the shape of the lever emerges. Excess brass (flashing) squeezes out around the edges of the forging and is trimmed on a punch press. The lever then goes from grinders to polishers to buffers and finally to the lacquer spray booth.

Quality and standards—Many companies produce a broad range of locksets that vary, not just in style and finish, but in quality. Two locksets made by the same company may look exactly alike behind the cellophane windows on their packages, but one might cost $25, and the other, $70.

Among the factors that determine the cost of a lockset are whether the key cylinder is made of turned brass (which can be machined to close tolerances and is long-wearing), a zinc die casting, or plastic. The steel parts inside a good grade of lockset are treated for corrosion resistance; cheaper ones aren't. One locksmith I talked to suggested picking up the locksets to gauge their weight. The heavier one, she told me, is probably better. Someone else suggested working the mechanism and trusting my instincts about which feels smoother.

It's likely that the $25 lockset is a tubular-style and the $70 lockset is a cylindrical. Following World War II, manufacturers developed the ability to install key cylinders in tubular-style locksets. But in general, tubular-style locksets don't meet the same standards that cylindrical locksets meet.

The Builder's Hardware Manufacturer's Association, Inc. (BHMA, 60 East 42nd St., New York, N. Y. 10165) sponsors a set of lockset standards published by the American National Standards Institute (ANSI) and certifies locksets according to those standards. Products from companies that subscribe to the standards (not everyone does) are selected at random and without notice and are tested for strength, performance and finish by an independent laboratory. The grading that results is rather involved, but a few examples will give you some idea of what goes on.

Bored locksets (ANSI A156.2) are certified in three grades. Grade 1 means that the lockset is suitable for heavy-duty commercial use. Grade 2 means it's suitable for light-duty commercial use. And grade 3 means it's suitable for residential use.

In one test, a lockset is attached to a door that's opened and closed by machine at a rate of 10 times per minute. A grade 1 lockset has to be operational after 600,000 cycles, grade 2 after 400,000 and grade 3 after 200,000. In another test, a 100-lb. weight is swung into a closed door from a given height to see if the latch bolt will bend or break, allowing the door to open.

Should you consider installing grade 2 locksets on a house? Not necessarily. Grade 2 hardware will stand more abuse than grade 3, but the difference might only be whether it takes five minutes to break into the house or three minutes. In either case, the weak link is most likely to be the door and the jamb, not the hardware. Grade 2 locksets will certainly be more durable than grade 3. But whether they will see enough use in a house for the difference to become apparent is another question.

Not all manufacturers participate in the certification program, and not every product of the participating manufacturers meets the minimum standards. To make matters worse, most companies don't advertise their certification on packaging. They do, however, include it in their catalogs, so if you're interested, get a catalog. Or better yet, contact the BHMA and for $2.50, get a copy of their "Directory of Certified Locks and Latches."

Reaching for the brass knob—Brass, an alloy composed of 60% to 70% copper mixed with lead and zinc, is the premier material for door hardware. It offers the right combination of strength, workability and corrosion resistance. According to builders and hardware dealers I talked to, some of the finest brass knobs and handles generally available are made by the Baldwin Hardware Corporation (841 Wyomissing Blvd., Reading, Penn. 19603). If you're converting the room above your garage into a rental unit to help meet your mortgage payments, Baldwin locksets and handles probably aren't for you. Their solid brass hardware is expensive, but considered worth the cost by those who can afford it. I visited Baldwin's plant to find out why.

The heart of Baldwin's operation is hot forging, the same process used by blacksmiths. Raw material, purchased from brass foundries in the form of bar stock, is cut to length, and the pieces, or billets, are conveyed into furnaces and heated to 1,400° F. The red-hot billets are then placed in dies and whomped into shape by a huge forging press. Excess material called flashing seeps out around the forging and is later trimmed on a punch press (photo, facing page).

Baldwin touts hot forging over the sand casting used by other companies by pointing out that the resulting pieces are denser, stronger and smoother, lacking pit holes and air pockets. But in the context of residential use, they aren't necessarily more durable or more secure. In this price range, the difference comes down to aesthetics—how it looks and feels, which is why Baldwin takes such pains with polishing, buffing and finishing.

The biggest challenge faced by Baldwin and by any manufacturer of expensive hardware lies in the nature of custom homebuilding. Door hardware is one of the last things to be installed in a house, and if the project is over budget, hardware is a common target for cost-cutting. To learn what a cost-cutting homeowner or builder might use instead of solid brass, I visited the Dexter Lock Company (300 Webster Road, Auburn, Ala. 36830). Dexter does sell a designer series of forged brass door hardware, but I wanted to learn how they make the ubiquitous hollow doorknob that most of us reach for every day.

At Dexter, doorknobs are formed on a sixteen-stage transfer press. The brass is 8 in. wide and .028 in. thick and is fed to the press from coils as big as wagon wheels. After cutting it into 8-in. squares, the machine transfers the brass from one die to the next, forming the familiar shape a little more each time and trimming off the excess. Toward the end of the process, the knob is filled with fluid and expanded by hydraulic pressure to create the final shape. The whole process takes less than a minute.

The big difference between solid brass and hollow brass doorknobs ends at that point. Both get much the same white-glove treatment as they're passed from polishers to buffers and then conveyed on racks through the lacquer spray booth and drying ovens.

Exotica and where to find it—The Schlage Lock Co. (2401 Bayshore Blvd., San Francisco, Calif. 94134) recently introduced something called a Key N' Keyless Lock. It's a lockset with optional dead bolt that can be opened without a key simply by twisting the knob left and right in sequence. A British company, Modric, Inc. (P.O. Box 146468, Chicago, Ill. 60614) makes lever handles in 355 colors, with cabinet hardware and bath accessories to match (photo below, center).

Valli & Colombo (P.O. Box 245, 1540 Highland Ave., Duarte, Calif. 91010), an Italian company, recently introduced a designer line of door handles developed for the disabled (photo below, left). Meroni, another Italian company, makes a push-button doorknob (distributed by Iseo Locks Inc., 2121 W. 60 St., Hialeah, Fla. 33016) that opens when you squeeze it (photo below, right).

Normbau Inc. (P. O. Box 979, 1040 Westgate Dr., Addison, Il. 60101) and Hewi, Inc. (7 Pearl Ct., Allendale, N. J. 07401) both make colorful nylon-coated door hardware that's tough and won't corrode or tarnish.

You probably won't find these products in local hardware stores or lumber yards. Look in the Yellow Pages under "architectural or builder's hardware," or go to a locksmith. If you are interested in antique or reproduction hardware, check out the listings in the *Old-House Journal Catalog* (Old-House Journal Corporation, 69A Seventh Ave., Brooklyn, N. Y. 11217. $15.95, softcover). You can also write to the Door and Hardware Institute (7711 Old Springhouse Road, McLean, Va. 22102-3474) for a copy of their *Buyer's Guide, 5th Edition*. It will cost you $35, but you'll get the most comprehensive list of door-hardware manufacturers that I've seen. □

Kevin Ireton is an associate editor with Fine Homebuilding.

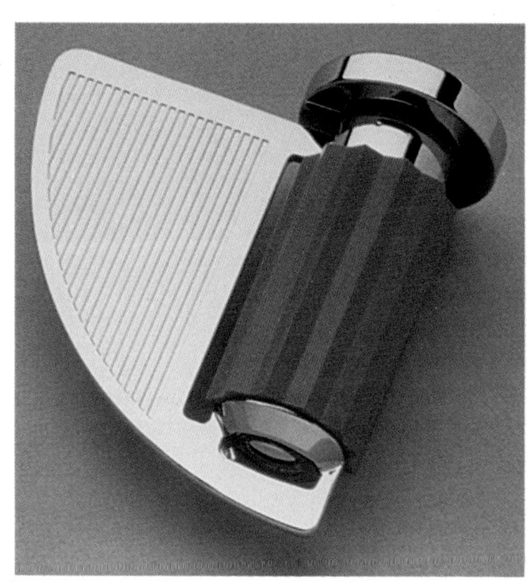

Designer handles for the disabled (left), made by Vali & Colombo, ergonomic levers in 355 colors from Modric (center) and push-button doorknobs from Meroni (right) are just a few of the European door handles recently introduced in the United States.

Batten Doors

Building a solid door from common lumber

by Bruce Gordon

A fine-looking batten door can be made from materials sold at any building-supply house, and can be built with limited funds and equipment. In the years when our business had no shop and little machinery, we produced custom batten doors at job sites, using only a table saw, an electric drill and a few clamps. They looked great, and were also competitive in price with factory-made doors.

Batten doors do have some inherent problems, though. Wood moves. A 36-in. door can vary as much as ⅜ in. in width between a dry winter and a humid summer. This will show on the side opposite the hinges, and the door that fits perfectly this winter may need to be planed down next summer and have its latch mortise reworked. The problem can be minimized by accommodating wood movement in the construction. Sealing the wood also helps, but if you use an oil finish, the door will move more than if you use varnish.

As a rule, batten doors do not stay perfectly flat and straight. They tend to bow across their widths and sag away from their hinges. The severity of these problems will depend on the species, grading, dryness and thickness of the lumber that you use, and how carefully you put the door together.

Boards 1 in. thick are best for interior doors, as are 1⅝-in. boards for exterior doors, although you can use ¾-in. tongue-and-groove stock for interior doors and 1½-in. stock for exterior doors. The batten should be 1¼ times as thick as the door body, and 6 in. to 8 in. wide.

To begin, select the stock and cut it to approximate size. If it is roughsawn, joint one face, thickness-plane, joint one edge, rip to width, cut to length, tongue or groove (or half-lap) each edge and do any decorative milling.

I edge-join the boards with dowels to keep the door from sagging away from its hinges and use a Stanley self-centering doweling jig to drill two holes for 2-in. long hardwood dowels in the mating edges of all the boards. The holes should be drilled level with the eventual position of the hinges—about 13 in. from the top of the door and 7 in. from its bottom. Dowel diameters vary with the thickness of the stock, but I use ⅜-in. dowels with ¾-in. boards, and ½-in. dowels with thicker stock. Once the holes are drilled, I apply a sealer (usually Watco or tung oil) and a first coat of finish or stain to the boards. On a door that will be painted, a coat of primer will do.

On a flat surface, assemble the door, inserting the hardwood dowels in the drilled holes. The dowels should not be glued, nor should the boards be pulled up tightly in the clamps. Instead, I insert strips of Formica between the edges to produce uniform gaps between the boards (photo facing page, top). The resulting gaps allow the wood to expand.

Being careful to keep the eventual location of hardware in mind, lay out the battens on the back of the door. Several pattern possibilities are shown in the drawing below. Check to be sure the door is square and flat, then clamp the battens in place. Battens should never be glued to the body. Attach them with metal fasteners so the wood can expand and contract freely.

I use four types of fasteners: rose-head clinch nails for their old-style look and ease of application, drywall screws for speed when I'm not concerned with the looks of the batten side of the door, wood screws, countersunk or counterbored and plugged, and carriage bolts for the substantial look they give the face of the door. In any case, the batten should have an oversized hole, to allow the body of the door to move without bowing the door. Be careful to predrill even for clinch nails to avoid splitting the wood where the nail breaks through on the opposite side. Ideally, the clinch nail should be bent twice so that it penetrates back into the wood (drawing, facing page), ensuring a tight fit. However, it is common practice simply to fold the clinch nail flat. When the battens are secure, remove the Formica spacer strips.

A few tips regarding hardware may help you avoid frustration. Interior batten doors are usually too thin to take either a mortised latchset or a cylinder latchset. Consequently, you should plan on either a thumblatch or a rim lock. Batten doors are most often installed with a strap hinge, an H-L hinge or an H hinge.

If you don't want the traditional look such hinges give a door, you can use butt hinges. They should be sized so that the screws fasten to the edge of the door itself, not to the end grain of the battens where they won't hold. There are also offset hinges that can help you work around batten placements. Size the hinge so that its throw and the length of the batten allow the door to open 180° without hitting the casing. In some instances it may be best to hang the door from the casing, not the jamb, or use a half-surface hinge, one that combines a strap across the surface of the door with a butt plate mortised into the jamb. □

Bruce Gordon is a partner in Shelter Associates, a design and building firm in Free Union, Va.

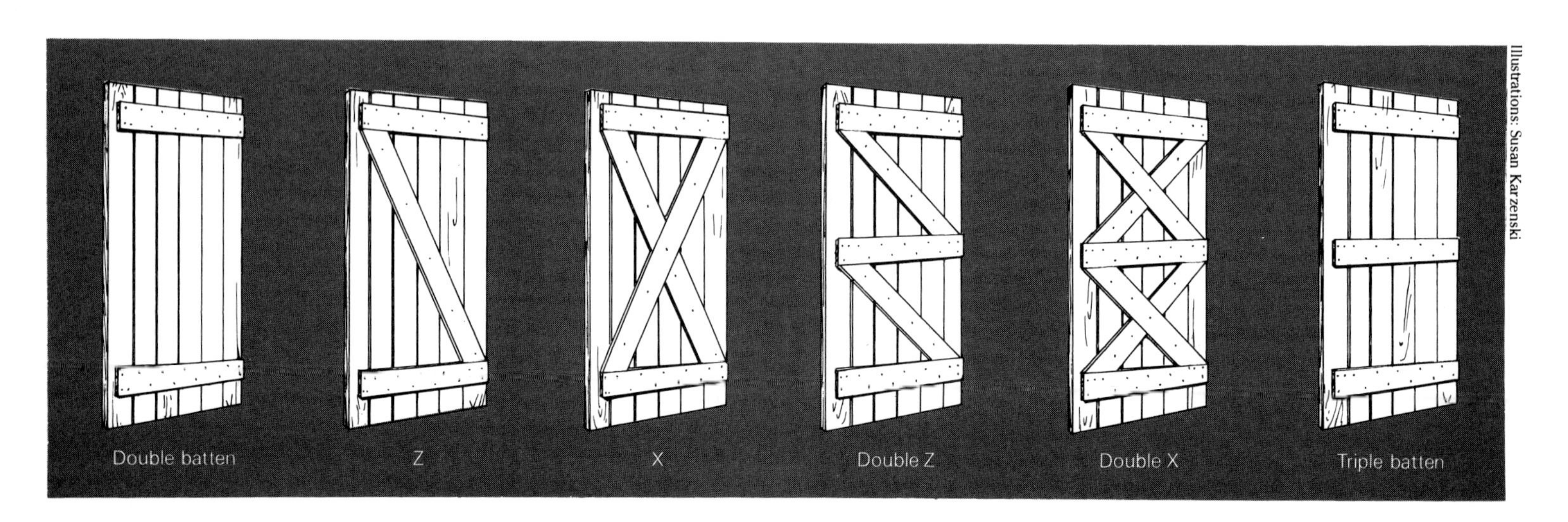

Illustrations: Susan Karzenski

From *Fine Homebuilding* magazine (February 1982) 7:32-33

Tongue-and-groove stock cut to approximate length is laid out across two sawhorses and clamped together flat and square. Two hardwood dowels inserted in holes in the mating edges of each board prevent sagging (detail, below). Above, battens are clamped to the back of the door. Formica strip spacers produce gaps that will permit the inevitable swelling. Battens should be fastened with either drywall screws, clinch nails or wood screws (left to right in bottom photo) or carriage bolts. Battens are predrilled for oversized holes, right, to allow for wood movement.

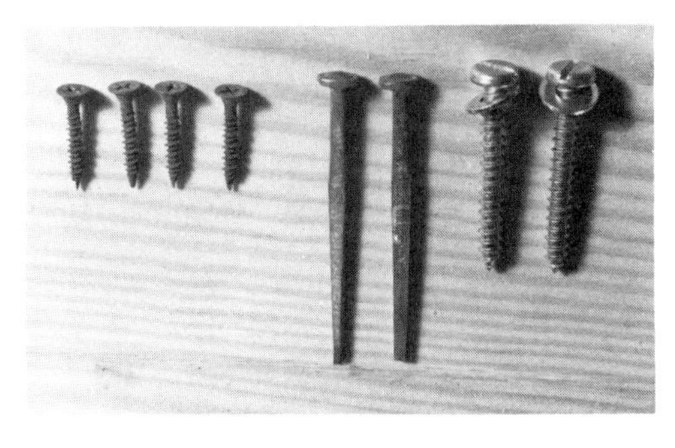

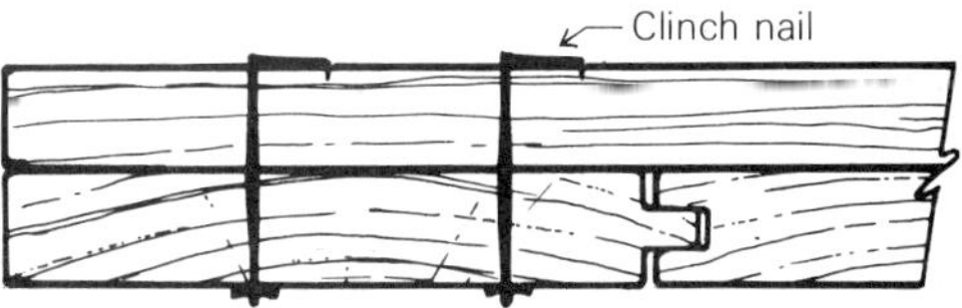

Photos: Susan Mortell

Curved Doors

A design with a circular closet in the center of the house calls for some special techniques

by Thomas Duffy

The circular storage unit

Several years ago, I was hired to redesign a small, single-story house to accommodate a family of three. The bungalow had been chopped up into a number of little rooms, and the new owners wanted more convenience and openness. It was obvious from the start that storage was going to be a big problem in this 24-ft. by 28-ft. space, and I wanted to make use of spots that might otherwise be wasted.

I began by dividing the floor plan into quarters and drawing in corner cabinets, as shown in the drawing, below right. Then I sketched in the bathroom and a short hall to get to it, a process that transformed the four corner cabinets into a single discrete structure at the center of the house. This pushed me out on a limb. Putting major storage dead center in a house plan isn't exactly a standard design solution. But it occurred to me to play with the idea of round storage space in place of the square closets or shelves, and I liked the result. With the rest of the house still laid out with straight walls, curves at the center would give it a sense of depth and spaciousness that might be surprising and delightful in such a small space.

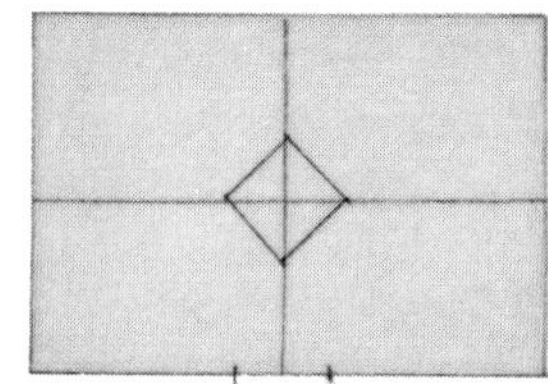
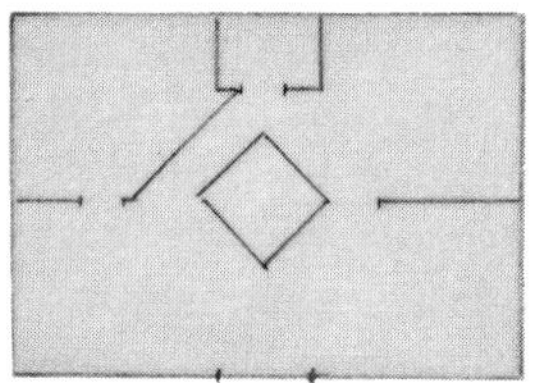
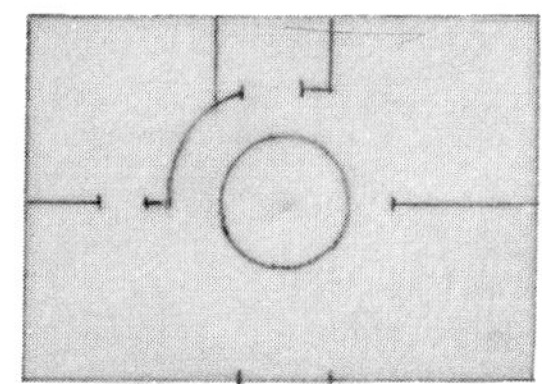

Evolving the design. *The author divided the bungalow into quadrants, with corner cabinets for storage (above left). Next, he sketched in the bathroom and a short hall (center). This left an unattractive freestanding closet in the center of the house. Duffy refined the idea of central storage to a circular closet and a curved hall (right).*

Getting down to design details, I decided that the closet would be 7 ft. in diameter, containing kitchen storage, a linen closet and a bedroom closet, a sliding door pocket and a bookcase. A short section of curved wall in the bathroom hall would be the arc of a 14-ft. dia. circle swung from the same center. I'd recently found a quantity of butternut at a good price. I had never worked with this wood, and was eager to give it a try. My clients liked the way it looked, so we decided that all of the woodwork in the house would be butternut. As it turned out, the wood worked beautifully, and hand tools left it with a lovely, bright surface.

My crew's first job was to remove the existing partitions. Then, using trammel points, we traced the curve for the rough frame onto 2x12s. We bandsawed our top and bottom plates from this stock, and installed the top plate first, dropping plumb lines to locate the position of the bottom plate. We toenailed in the radial 2x4 studs of the closet's exterior walls, and installed door headers and cripples.

The door pocket was framed up out of 2x2s. Because of the curving surfaces, we decided to use plaster instead of drywall, so once the framing was done, we installed plaster grounds. These are 5-in. wide strips of wood that are nailed around the rough openings to be flush with the finished plaster surface. They look somewhat like casings, but are meant to act as guides or screeds while the wall is being plastered. They were removed and replaced with the actual casing after the plasterers and painters had finished their work.

Next, I installed the 4x4 center post, toenailing it to the subfloor below and to a nailer between the joists above. If the center post were taken out of the closet today, it might look like a crisp piece of sculpture, with all of the mortises, rabbets, notches and chamfers that were cut into it to accommodate the shelves, partitions and nailers.

I used interlocking joinery for two reasons. The first is philosophical: I like to make a piece of wood do as much as it can. I'd rather cut mortises than apply more wood to do the job. The second reason is practical: Unglued interlocking joints—dovetails, for example—can be taken apart and put back together again if any adjustments have to be made as the project unfolds.

I mortised horizontal shelf supports of 5/4 pine into the post, and toenailed their other ends to framing members. Then I nailed up vertical tongue-and-groove pine partitions between storage areas, and screwed cleats to the

From *Fine Homebuilding* magazine (December 1982) 12:61-63
Photos: Allen H. Burns; Illustrations: Barbara Smolover

partitions for the shelves, parts of which also had to be mated with the central post.

Making these doors was surprisingly easy. The three rails are the only bent sections, and the only differences from straightforward door-building are that you have to bend the rails, feather the flat panels slightly and adjust the layout so you can worry the joints between rails and stiles closed. These operations are shown in the drawings on the next page.

The top and center rails of these doors are 5 in. wide, while the bottom rail is 7 in. wide. I made each rail of three ½-in. thick plies of butternut, each of which was steamed for an hour in a box and then bent over a form that I glued up out of boards cut to the proper curve. I used winding sticks to make sure that the form curved smoothly without any twist (drawing A).

I left the middle ply a couple of inches long on each end so I would have ready-made tenons. I also cut the center ply ½ in. narrower than the outside plies on the top and bottom rails. When the rails are glued in place, its narrow width forms a groove to hold the panels. The center rail needs grooves to accept panels on both sides, so I left the middle ply a full inch narrower than the outer ones. In the center of each middle ply, I also cut a notch 1½ in. deep and big enough to accept a stile. When the rails were glued up, this notch became a mortise for the center stile (drawing B). The butternut was roughsawn, so I cleaned up the field of the panels and roughed out the bevels with a scrub plane, a plane with a convex iron that will remove a lot of stock very fast. Then I jack-planed the field, which left a lovely surface, and finished raising the panels with a panel plane. The doors' stiles were cut slightly thicker and wider than their finished dimensions would be so that I could properly locate their grooves and then plane them to conform to the curve of the rails, and to fit the jamb exactly. I built the jambs in my shop and prehung the doors there.

More curves—There were two other curved elements in the woodwork I did for the house: the concave corner cabinet in the kitchen and the circular bathroom vanity (photos at left). Both of them required tighter bends than the doors, and no amount of feathering would have made flat panels fit. I had to fabricate curved panels as well as curved rails. To do this, I built a cradle out of ¾-in. plywood and 2x2s that established the right curvature (drawing C). Then I laid up 1x3 strips of butternut, after determining the appropriate bevel angle for their edges, as the drawing shows. I cut the bevels on the table saw. The bevel on each strip needed some work with a plane so that it would fit tightly next to its neighbor, stave-fashion. Once the curve looked smooth, I glued the strips together. This work took a lot of time and patience, and I did a lot of dry-fitting before I got everything to mate. I cleaned up the surfaces of the curved panels with scrub and compass planes, and raised the panels, which were still in the cradle, with the panel plane along the grain, and a small-radius gouge and chisels on the curved ends.

More curved built-ins. **The curve became a unifying element in the remodeled bungalow. The kitchen corner cabinet (above left) and the bathroom vanity (left) are variations on the theme. Unlike the doors of the large circular closet, which have curved rails and flat panels, the corner cabinet and vanity have curved panels as well, because of their small radius.**

Making and joining the rails and panels. **The rails of the curved doors are glued up from three ½-in. plies, steamed and bent over a form (drawing A, next page). The middle plies are cut long, leaving ready-made tenons for joining the rails to the stiles. They are also ripped narrower than the outer plies, and are positioned to form grooves for the panels to ride in. The middle ply of the center rail is ripped extra narrow to accommodate both top and bottom panels. Notches in the middle plies create mortises for the center stiles (B).**

Gently curving doors don't require curved panels. Flat panels are simply feathered to fit their grooves. The curved panels of tightly curving doors were assembled on a specially fabricated cradle (C).

While I was working on this house with its beautiful wood and admittedly unusual design, I felt a strong pull toward what I call quirky work, the eccentric tour de force that all too often has no practical purpose. My temptations led to some reflection on the role of craftsmanship. I began to take a fresh look at my ideas about moldings. I have no formal training in either design or woodworking, but it was clear to me that changes in plan or elevation are not to be taken lightly. I began to wonder about all this graceful static that Chippendale and his friends were so fond of. Was there something to hide? Did the wealthy need a performance every time a horizontal met a vertical? Did they need to avoid recognition of a change? What is this blur in the corners?

I concluded that the best work shows tasteful restraint. Consider the Gamble house by Greene and Greene. For my money, its magnificence is diminished by the extreme amount of thought and action devoted to every feature. No surface was left alone, no plane unaltered. Even rafter ends and door stiles make powerful statements. The house is loved for its honesty, and rightly so, but it is a bullying honesty. I decided on subtle, spare decoration where I wanted any decoration at all. For instance, along the edges of the doors' stiles and rails, I cut a simple bead, which I think makes each panel look less like a picture in a frame, and also results in a gentle play of light and shadow that adds to the warmth of the house.

This job also taught me that you don't need vast spaces to create an interesting design. In fact, I think that a small room or house offers greater possibilities for good design, because there's no space to waste. By introducing angles or curves and by paying attention to details and craftsmanship, you can create a jewel-like richness and a space that feels much larger than it really is. I'm happy that the very center of this little house turned out to be the best place for storage. □

Tom Duffy is an architectural woodworker who lives in Ogdensburg, N.Y.

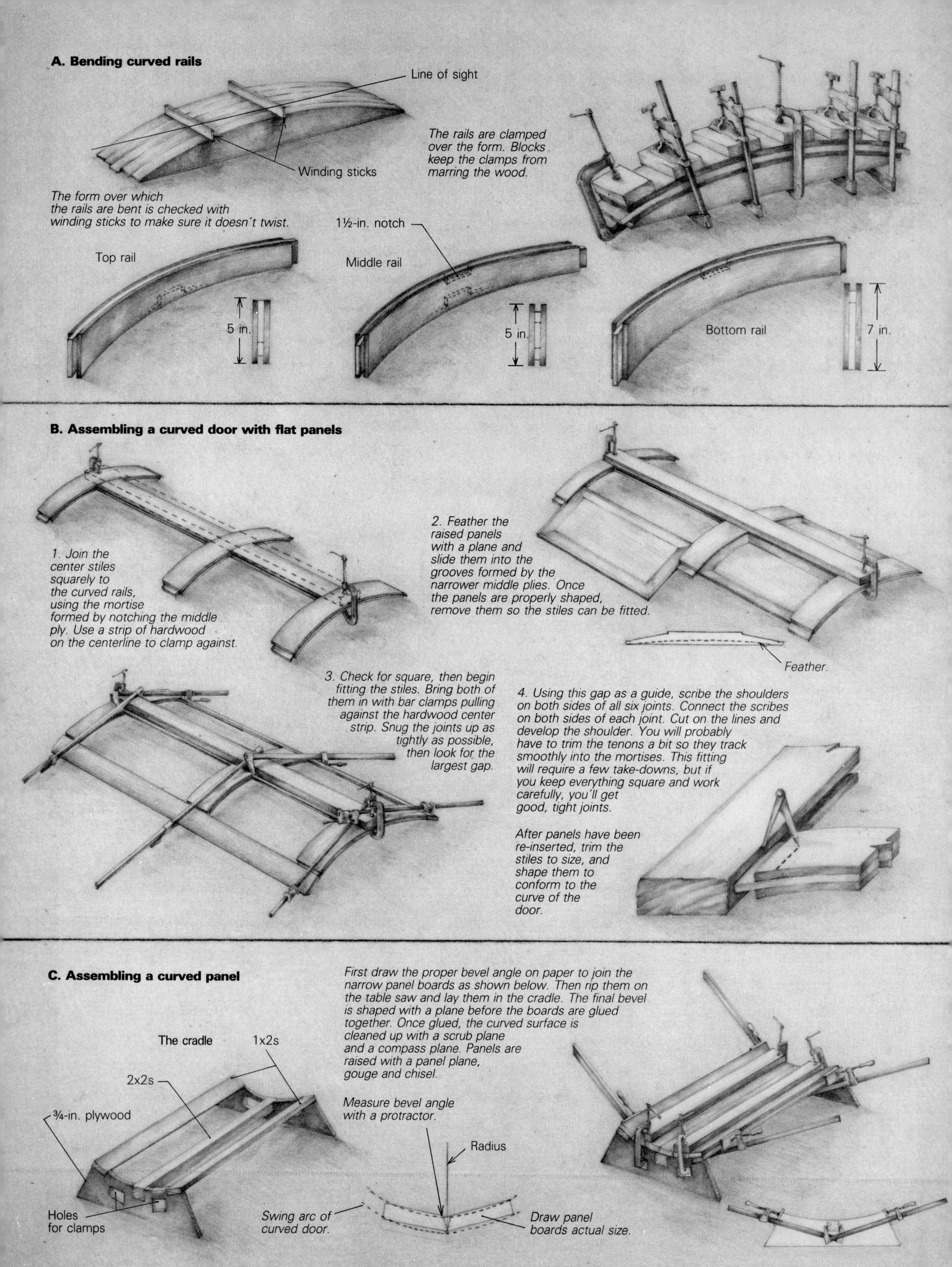
A. Bending curved rails
Line of sight
Winding sticks
The form over which the rails are bent is checked with winding sticks to make sure it doesn't twist.
The rails are clamped over the form. Blocks keep the clamps from marring the wood.
1½-in. notch
Top rail
5 in.
Middle rail
5 in.
Bottom rail
7 in.
B. Assembling a curved door with flat panels
1. Join the center stiles squarely to the curved rails, using the mortise formed by notching the middle ply. Use a strip of hardwood on the centerline to clamp against.
2. Feather the raised panels with a plane and slide them into the grooves formed by the narrower middle plies. Once the panels are properly shaped, remove them so the stiles can be fitted.
Feather.
3. Check for square, then begin fitting the stiles. Bring both of them in with bar clamps pulling against the hardwood center strip. Snug the joints up as tightly as possible, then look for the largest gap.
4. Using this gap as a guide, scribe the shoulders on both sides of all six joints. Connect the scribes on both sides of each joint. Cut on the lines and develop the shoulder. You will probably have to trim the tenons a bit so they track smoothly into the mortises. This fitting will require a few take-downs, but if you keep everything square and work carefully, you'll get good, tight joints.
After panels have been re-inserted, trim the stiles to size, and shape them to conform to the curve of the door.
C. Assembling a curved panel
First draw the proper bevel angle on paper to join the narrow panel boards as shown below. Then rip them on the table saw and lay them in the cradle. The final bevel is shaped with a plane before the boards are glued together. Once glued, the curved surface is cleaned up with a scrub plane and a compass plane. Panels are raised with a panel plane, gouge and chisel.
The cradle
1x2s
2x2s
¾-in. plywood
Holes for clamps
Measure bevel angle with a protractor.
Radius
Swing arc of curved door.
Draw panel boards actual size.

A Breath of Fresh Air

Making a Victorian screen door

by Alasdair G. B. Wallace

On a trip through southern Ontario, I was impressed by the number of carefully restored and lovingly maintained older homes. Wherever I looked, there were beautiful old doors—some were hand-carved, many swung on wrought-iron hinges, others had leaded lites. But it saddened me to see these original doors hiding behind standard aluminum combination storm-and-screen doors that lacked aesthetic rapport with their surroundings. Southern Ontario does require storm doors and screen doors in the appropriate seasons. Antique doors occasionally show up at country auctions, but they usually command exorbitant prices. Making such a door, however, requires only basic woodworking skills, and will provide you with a good excuse for an indoor project this winter and a welcome breeze next summer as well.

The door detailed in this article is a copy of a well-worn original. It consists of an inner oval frame enhanced by turned spindles, a separate lower decorative screened opening, and a raised panel at the base (photo right). The frame of the door may be joined by dowels (as was the original) or mortised and tenoned. I recommend the latter for its greater strength. You'll require a table saw, basic woodworking hand tools and a lathe. A bandsaw or saber saw will speed some of the work. Some of the cuts and grooves described below can also be made with a radial-arm saw, a shaper or a router.

Materials and hardware—Regardless of the wood you use—oak and pine are popular in Ontario—you'll need boards that dress out to a minimum thickness of ⅞ in.; 1 in. is preferable. Anything less will result in a flimsy door. Select wood that is dry (9% moisture content or less), straight grained, and quartersawn if possible. You'll need about 18 bd. ft. of stock, and about 2 bd. ft. of 4/4 maple for the spindles.

For my door I purchased new hardware from Lee Valley Tools Ltd. (2680 Queensview Drive, Ottawa, Ont., Canada K2B 8H6). You might also take a look at their catalog of antique hardware. You'll need one 9-in. Chicago door spring with an adjustable tensioning device and a cast-iron screen-door set, which includes a mortise lock, knob, drop handle and strike plates. Make sure it is suitable for doors at least 1 in. thick. Brass sets are also available. Also get three standard butt hinges. Since I planned on alternating the screen door with a standard storm door as the seasons changed, the hinge leaves for each are identical in size and placement, and align with the leaves on the door frame. I have only to remove the pins to change doors.

For screening, you'll need enough to allow a 2-in. overlap on all sides of the opening. Bronze and copper screening (if you are able to obtain them) are strong, an important consideration if you have children or a pet or are plagued with squirrels. I used black anodized aluminum screening. It is easily installed and becomes almost invisible against the shadows.

Laying out the plans—Don't assume that your door frame is either square or plumb. Measure from corner to corner diagonally. If the diagonals are the same, your frame is true; otherwise you'll have to make allowances in the door. Measure the width and height at several locations, then make your screen door ¼ in. shorter and ⅛ in. narrower than the opening. In calculating the length of your rails, remember that you need to add 3 in. for the 1½-in. long tenons (drawing, next page).

Making the door—Select the straightest-grained boards for the stiles. Once you have marked out your stock, rip it to the required dimensions and square the edges. I prefer to remove any planer marks with a sharp smoothing plane because it leaves the surface of the wood bright and crisp-looking. If you wish, though, you can sand them out. To avoid possible error later, lightly pencil in the future location of each piece—top rail, left stile—on its outdoor face.

The design of this door includes a decorative molding, which may readily be cut on the rails and stiles with a shaper head such as a Craftsman 9-2352 AM or a Multi-plane. I prefer the Stanley 45

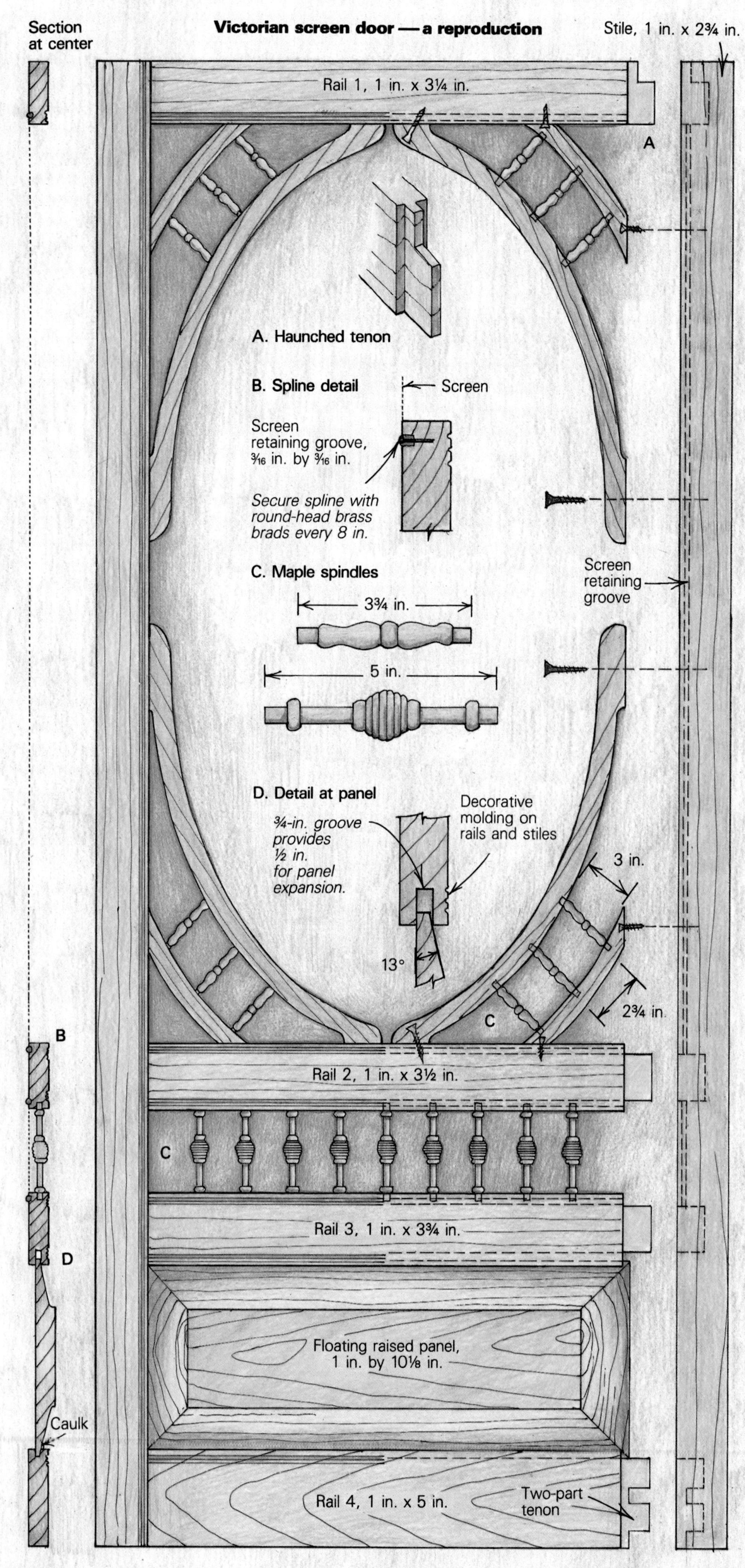

grooving plane, but this old tool is hard to come by. A series of passes over the angled blade of a table-saw will achieve a similar effect, or you could use a router. But to avoid weakening either the mortises or the groove for the panel, run the pattern to a depth of only 1/16 in.

The tenons should be approximately one-third the thickness of, and centered in, the stock. For additional strength, tenons in the top and bottom rails should be haunched (see detail A at left). Because the bottom rail is 5 in. wide, its tenon should be in two parts in order to reduce the possibility of weakening the stile.

Once the tenons have been marked out, they may be cut in the traditional manner with a backsaw. An alternative method uses the simple tenon-cutting jig described in the sidebar on the facing page.

Mortises can be simply marked out by laying the rails across the stiles in the appropriate location and marking the edge of the stock around the already cut tenons. Mortises may be cut with a router, but you can also chop them out with a mortising chisel. If you use a router, either the edges of the tenons must be rounded over with a rasp to match the rounded corners of the mortise, or the mortise must be chiseled square to match the tenons. Either way, accuracy is essential if you want to avoid having the door twist out of shape.

Run the grooves for the lower panel in the noted locations in rails and stiles. Making these grooves the same thickness as the tenons and centering them on the edge of the stock enables you to run them through to the ends of the stiles, where they serve to locate the haunched part of the tenons. If you're using the dado head on a table saw, be careful not to over-run. Mark the location of the blind end of the groove in pencil on the face of the stiles. Mark the location of the leading edge of the dado cutter on the rip fence, stopping the groove just before the two marks align. Square up the stopped ends with a chisel.

After you have dry-assembled the frame of your door to check dimensions and flatness, mark on the inside face of the door the location of the 3/16-in. by 3/16-in. groove that will house the screen-retaining splines. This groove should be ¼ in. from the inner edge of rails 1, 2, and 3 and the stiles. Running these grooves the length of the stiles will considerably weaken the door in the area of the mortises and the panel-retaining groove. Instead, I recommend that you match the ends of these grooves with stopped grooves in the stiles. A router is a most convenient tool for this operation.

To make the panel, you'll have to glue up several narrower boards. Make sure you alternate the direction of the annual rings in order to minimize the liability of warping. Rather than shape both faces of the panel, I decided to leave the back flat. This results in the front face protruding beyond the rails and stiles, adding a bolder relief to the bottom part of the door.

Once the panel blank has been dimensioned, the center raised portion can be formed on the table saw. To do this, tilt the blade to an angle of about 13° and run the panel through on edge, using a 6-in. high rip fence to help support the

Drawings: Frances Ashforth

work. With a sharp carbide-tipped rip blade, you can raise the panels in a single pass. To cut the small vertical shoulders that border on the field, lay the panel face down and use the fence as a guide. Experiment with scrap wood to achieve the effect you want. Clean up the bevels with a rabbet plane and sandpaper block. The finished panel edges should be ¼ in. thick.

The width across the grain of a 10⅛-in. panel may vary considerably, depending on the species of wood and the ambient humidity. To accommodate this potential movement, run the panel-retaining groove in the bottom of rail 4 to a depth of ¼ in., and allow about ¾ in. for the groove in rail 3. The panel will be fully seated in the first groove, and the second will accommodate expansion.

Spindles and inserts—I experimented with several shapes and spacings of spindles before deciding on the combination that looked best. I used nine of the smaller spindles and twelve of the longer ones (detail C on the facing page). Mark the locations of the spindles on the bottom of rail 2 and the top of rail 3, and center the holes in the stock unless your frame is less than 1 in. thick. In that case, offset the location of the holes toward the outer face of the door so that they don't run through to the screen-retaining groove on the inner face of these rails.

The inserts that define the oval shape of the door's interior also serve to strengthen the frame. Prepare two full-size patterns for the two different parts of the oval frame insert, and lay out the stock so that the grain runs with the length of the stock. After the stock has been marked, cut out the inserts with a bandsaw or saber saw. On my door, I rounded the inner edges slightly with a ¼-in. cove bit fitted to my router. Use a framing square to make sure the ends are square to each other. The inspiration for this screen door had inserts that were nailed in place, but I prefer to fasten the inserts with 2-in. #9 brass screws, countersunk.

Once the entire door (with the exception of the oval inserts and spindles) has been dry assembled, test-fit it in the door frame. Note and correct any irregularities before beginning the finishing process.

Finishing and assembly—Traditionally, many pine doors were painted, while others were finished with a varnish stain and still others were grained to simulate oak. The door I built was lightly stained to match the existing pine exterior door, then finished with four coats of exterior satin Varathane. I rubbed each coat down with steel wool, using progressively finer grades down to #000.

I learned the hard way on the first door I built that finishing is much easier if it is done before the frame is glued up. This makes it easier to get at all the nooks and crannies of the completed door, and later, when the raised panel shrinks during dry weather, no unfinished margins will appear. Finish the spindles as they rotate on the lathe; then cut them to their final length. Be careful, however, to keep finish away from any portion of the door that will receive glue.

Gluing up will proceed more smoothly and rapidly if you have a helper to assist in aligning the panel and the lower spindles. Check that the frame is flat and square. If you need to adjust it, a long clamp placed diagonally from corner to corner will enable you to squeeze it slightly into square. (An additional advantage of finishing the parts of the door prior to assembly is that any excess glue can be readily removed from the finished surface.)

Don't glue the panel in place. Instead, leave it to float within the grooves in the surrounding frame. If your door will be exposed to the weather, run a thin bead of clear silicone caulking along the exterior seam between panel and bottom rail to keep water from collecting in the groove. Once the glue has set (remember to use a waterproof plastic-resin glue), the oval inserts and spindles may be installed.

Installing the screens—The next step is cutting enough 3⁄16-in. by ¼-in. pine splines to retain the two separate screens of the door. I gently rounded the top edge and tapered the sides of my splines with a block plane. The taper accommodates the double thickness of screening within the groove, and the pine itself is soft enough to compress slightly in the groove.

You don't need anything fancy to persuade the screening into place. I used a scrap of metal plate and lightly tapped it with a mallet to push the screen into the grooves, long sides first. Enlist your helper to stretch the screening as you urge it into the grooves, checking frequently that it is taut. For a good-looking job, make sure that the weave of the screen aligns with the frame. Cut your splines to length and tap them into position, using a scrap of wood to avoid scarring them. Secure the splines every 8 in. with ¾-in. brass escutcheon pins. Any excess screening along the edges can be removed with a razor knife run along the outer edge of the spline—I trim it flush with the door. Finish the splines to match the door before installing them.

Installing hardware—Test-fit your door and mark the location of hinges and the handle. Then install them and hang the door. On my door, I installed a Chicago door-closing spring with adjustable tension. To eliminate most of the racket when the door bangs shut, I slightly recessed three 2-in. lengths of self-adhesive weatherstripping in the door frame.

I'm a slow worker, so this door and another one I made took me quite a while. But the breeze from the verandah through the kitchen area and the open feeling on the back porch make all the work worthwhile. My neighbor stopped me as I was mowing the lawn this morning, asking if I might find time to make a door for him. But I think I'll try designing one for our front door first, in maple, maybe, to match the staircase. □

Alasdair G. B. Wallace, of Lakefield, Ont., is a contributing editor of Fine Homebuilding. *Photo by the author. For more on this subject, see Amy Zaffarano Rowland's* Handcrafted Doors and Windows *(Rodale Press Inc., 33 E. Minor St., Emmaus, Pa. 18049, 1982; $12.95 paperback; $21.95 hardcover).*

A useful tenoning jig

If you have a series of tenons to cut, this tenoning jig is simple, inexpensive and will save you a lot of time—with one pass over your table saw you can cut both cheeks of a tenon (drawing, below). The idea is not mine, but I've been using it for years and am indebted to whoever figured it out.

You'll need two identical blades. I use 9-in. planer blades, running side by side in my 10-in. table saw. I use these blades only for tenoning, and sharpen them together in order to keep their diameters identical. They are separated on the saw's arbor by a metal spacer. A local machine shop made me a set of spacers, ¼ in., 5⁄16 in., ⅜ in., and ½ in. wide, and they have saved me endless hours and plenty of frustration. Check your saw's throat-plate clearance before you order your blades. To accommodate the width of the paired blades, I had to make a special table plate out of plywood. For safety, the throat-plate opening should be only slightly larger than needed to clear the blades.

The jig itself consists of a U-shaped wooden device which fits over the rip fence and slides along it. The fit should allow the jig to slide freely but not be so sloppy that

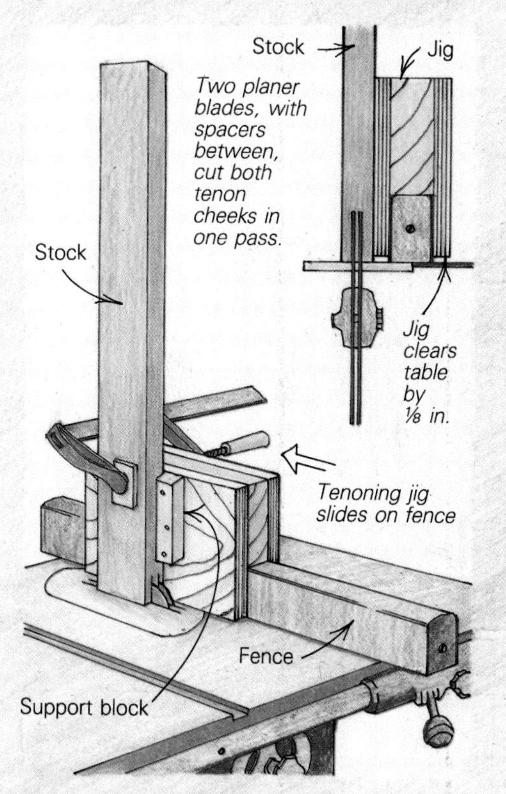

the workpiece wobbles. To minimize binding, you might try waxing the inside of the jig, or the fence. To cut the tenons, install the appropriate spacer between the blades so they will cut the exact thickness of tenon you wish. Raise the blades a distance equal to the length of the tenon, and adjust the fence with its sliding jig in order to center the tenon in the stock. Align the stock against the vertical backing block, clamp it in place, run it through the blades and you have two perfect cuts in one safe, efficient operation.

To remove the waste and cut the shoulders, make up an end-stopping attachment for your saw's miter gauge. Set your miter gauge at exactly 90° to the saw blade, and use a single planer blade to remove the waste. —*A. G. B. W.*

Making an Insulated Door

This handsome entry keeps heat in while it keeps breezes out

by Irwin L. Post

Well-insulated, tightly fitting doors reduce fuel bills and make a house more comfortable. I recently built one for a client in Weston, Vt., who wanted a weathertight door that would complement her post-and-beam house. Her interior doors are made from knotty eastern white pine tongue-and-groove paneling and have custom-forged strap hinges. We decided that the inside of the new door should have the same appearance. For the exterior, we settled on five vertical boards surrounded by a border, all of knotty pine (photo right). The hinges, latch and knocker were made by C. Leigh Morrell at the West Village Forge in West Brattleboro, Vt. For a door this thick, you'll either have to modify standard locksets or special-order commercial models. I silver-soldered an extension to the rod that turns the deadbolt.

The heart of the door is a core of pine spacers and 1-in. thick beadboard polystyrene foam (R-value: 3.3), sandwiched between two pieces of ½-in. plywood. The plywood facing stiffens the core and makes it dimensionally stable and resistant to warping. One advantage of using a structural plywood core is that it allows great freedom in designing the finished surfaces on both sides of the door. You can glue stock of any shape or thickness to the plywood skin, creating, for example, freeform or geometric patterns. The facing boards on both sides of the door are splined together. As the drawing shows, the exterior face is made from nominal 1-in. paneling, bordered with 1x stiles and rails. I applied ¾-in. trim strips to the vertical edges, and these are overlapped by the facing materials on both sides. I chose to make the core height equal to the full height of the door and not to cover the top and bottom edges of the core with trim strips, since they are not visible.

Irwin Post

Construction sequence—First, determine the size of the door. I made mine ¼ in. shorter and 3/16 in. narrower than the finished opening. Unless the door fits the opening precisely, air infiltration will render any amount of core insulation useless. Check to see that the opening is square. You can do this by measuring the diagonals—if they are the same length, then the opening is square. If it's not, you can either make the door fit the irregular opening or rework the jambs and casing to square things up.

Next, decide on the design for the door's inside and outside faces. It helps to make scale drawings to check the overall proportions of the various elements. Knowing the finished dimensions of the door, use your drawings to calculate the dimensions of the beadboard core.

Now, cut the two pieces of ½-in. plywood to the same dimensions as the core (the full height of the opening less ¼ in., the width of the opening less 1 11/16 in.). I used underlayment grade A-D ply with exterior glue. Cut 1x3 pine for spacers as shown in the drawing, step 1. For 1-in. beadboard, the spacers need to be a full 1 in. thick. Edge spacers can be made from short pieces if necessary, but they should be mitered and tightly butted to prevent air leaks. The diagonal spacers keep the door from twisting. The small blocks along the long sides are for extra reinforcement around the lockset. I glued a block on each side so I didn't have to keep track of which side would have the lockset and handle.

Glue the spacers to one piece of plywood. I used yellow glue. Be sure your worktable is absolutely flat, because a door built on a warped surface will be a warped or twisted door. A few 1¼-in. brads will keep the spacers from shifting while you apply clamps.

With a sharp knife or a razor, cut the beadboard to fit tightly between the spacers. Be sure not to leave any gaps. Now glue the second piece of plywood onto the spacers to complete the core (step 2). Clamps on the edges and cauls (cambered strips of wood which, when clamped at their edges, will exert downward pressure along their lengths) across the width of the door are a good way to get a bond. Use brads to keep the plywood from shifting as it's being clamped.

Attach the trim strips to the vertical edges (step 3). On this door, the strips were ¾-in. thick pine about 2¼ in. wide. I glued them on one at a time, and used a 2x4 to distribute the clamping pressure, as shown in the drawing. After both edges are glued on, trim them flush with the plywood on both sides. I used a ball-bearing pilot flush-cutting bit in my router, but a hand plane would work just as well.

While the glue is setting, prepare the paneling for the faces. I used splines and grooves to join adjacent knotty pine boards (step 4). You can use a dado head in a table saw to plow the grooves, or a slotting cutter in your router. The splines were cut on the table saw. The V-grooves that show on the outside of the door were made by beveling the boards with a hand plane.

Glue the paneling to the plywood core. I used no nails or screws because I wanted to fasten the panels in a way that would keep the gaps between the boards constant and minimize any distortion, such as cupping, due to seasonal changes in moisture content. To do this, I glued the boards along the outer edges of the door on their outer edges, and the others with a narrow bead of glue along their centers. I glued one board at a time, starting from one edge, and I used cardboard spacers between the boards during assembly to create space for expansion. The unglued splines allow for movement.

The door is most easily finished with it lying flat on a table. I stained the outside of this door, and then gave all exposed surfaces three coats of polyurethane varnish.

This door is 3½ in. thick overall. Its feeling of mass and sturdiness goes well with the post-and-beam construction. The total R-value is 7.25, calculated by the methods in the USDA Forest Products Lab's *Wood Handbook* ($10 from the U.S. Government Printing Office, Washington, D.C. 20402). By comparison, a standard 1¾-in. solid-core exterior door has an R-rating of about 2.5, and double-glazed windows have an R-value of about 1.9. ☐

Irwin Post is a forest engineer. He lives in Barnard, Vt.

From *Fine Homebuilding* magazine (August 1982) 10:46-47

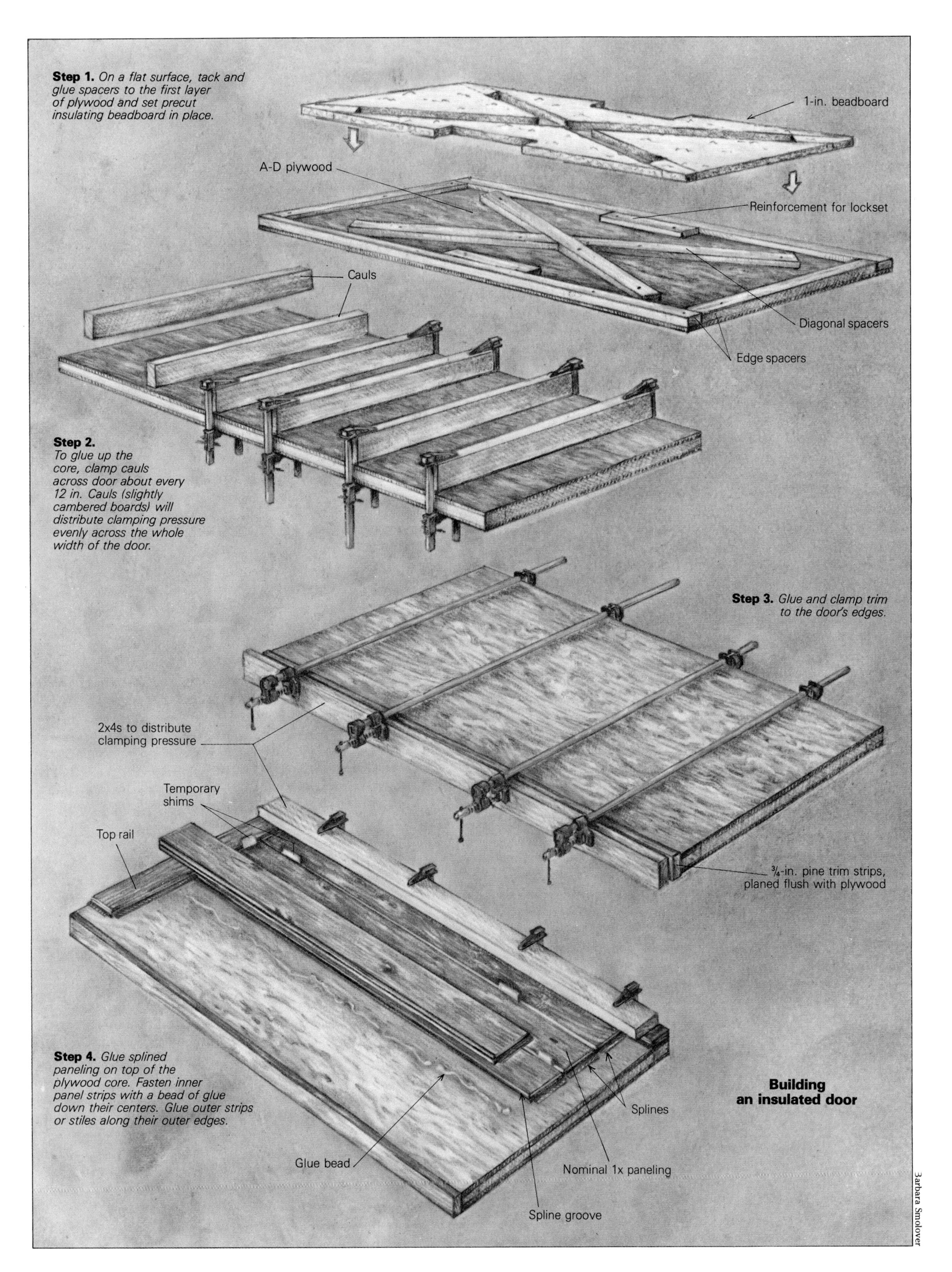

Step 1. *On a flat surface, tack and glue spacers to the first layer of plywood and set precut insulating beadboard in place.*
1-in. beadboard
A-D plywood
Reinforcement for lockset
Cauls
Diagonal spacers
Edge spacers
Step 2. *To glue up the core, clamp cauls across door about every 12 in. Cauls (slightly cambered boards) will distribute clamping pressure evenly across the whole width of the door.*
Step 3. *Glue and clamp trim to the door's edges.*
2x4s to distribute clamping pressure
Temporary shims
Top rail
¾-in. pine trim strips, planed flush with plywood
Step 4. *Glue splined paneling on top of the plywood core. Fasten inner panel strips with a bead of glue down their centers. Glue outer strips or stiles along their outer edges.*
Building an insulated door
Splines
Glue bead
Nominal 1x paneling
Spline groove
Barbara Smolover

Mudrooms

by Tim Matson

Lately, I've become a connoisseur of mudrooms, those traditional enclosed entryways now largely overlooked by architects and overshadowed by skylights, roof-panel heaters and solar greenhouses. As you can see by these photos, I've been scouting for examples of mudroom design to add to my mudroom bestiary, a collection of ideas that I plan to consult when I get around to adding on a mudroom of my own. I say bestiary because the mudroom embodies a unique architectural species. Rarely will you find two that are alike. Some are plain homemade, some ornamental, some solid and utilitarian. Often the designs say something about the people behind the door.

At its simplest, the mudroom is an entryway, usually added on to the outside of the house, where you can stomp the dirt or snow off your boots, or remove them, and hang up coat and hat before entering. More sophisticated mudrooms are fully enclosed, and even insulated, to combine the advantages of the shakedown entryway with an airlock for conserving heat. Other facets of mudroom design can produce payoffs in refrigeration, storage space, shelter for critters, and so forth.

The mudroom appears to have humble farmhouse origins, coinciding with the elimination of earthen floors. In Japan, where it's customary to slip off your shoes before entering a dwelling, the mudroom is known as the *genkan;* it's been a cornerstone of Japanese architecture since the 12th century. In Vermont, where I call home, the mudroom has been a popular architectural tradition since the first settlers kicked off their boots. It has sheltered the thresholds of taverns, inns, schoolhouses, stores, farmhouses and village homes.

The newest style of mudroom in my neighborhood is a portable entry to the general store. It's a three-piece modular unit—roof, walls and door—that gets assembled in November and taken away in May. "It's great," the storekeeper told me. "Keeps out the snow and keeps in the heat. The cashiers were freezing."

This roving mudroom is rare. Most set up house for keeps, and it's possible to learn a lot about the standard-breed mudroom in a one-room schoolhouse. There can't have been many places more in need of one. Send a dozen or more kids through the same door five days a week, nine months a year, and the advantages of the mudroom become clear. In one nearby village I came upon a century-old crimson schoolhouse recently converted into a community center, complete with a solar greenhouse on the south wall. Jutting out from the same wall stands the mudroom, which was wisely left intact. It is roughly 10 ft. by 10 ft. and capped with a hip roof. The floor is a thick concrete slab, perhaps the ultimate underlayment for withstanding the march of the Vermont seasons. There are benches for unlacing boots, pegs to hang wet clothes on, and a couple of large windows to provide natural light.

Another good lesson lies in the schoolhouse elevation, two steps above mudroom level. In winter, when the schoolroom door is open to the mudroom, heat resists spilling out.

The only thing wrong with this mudroom is that it has a hip roof, a design that creates an eave line on all of its three sides. Apart from a prejudice I have against hip roofs, as a mudroom roof it's preposterous. The watershed pattern democratically drenches all eaves, including and especially the one you must pass under to enter. Gutters are one solution, I suppose, but I'm accustomed to seeing roof ice twist them around like Budweiser pop tops. I'd rather vote for a peaked roof with the gable over the mudroom door.

At the opposite end of the scale you'll find the micro mudroom. It tends to be a homemade add-on, often not much bigger than a telephone booth. (Any day now I'm expecting to find a mudroom-sized phone booth planted at the front door of a glassy solar dwelling; what better way to recycle all those old booths displaced by modern plastic hoods?) Built on a foundation of posts, stones or concrete piers, the micro mudroom may not be as durable or spacious as the one sitting on a concrete slab, but then it's more manageable to finance and construct. After all, it *is* just a mudroom. Stud-wall wood-frame construction is the norm, as is the lack of insulation, unless the mudroom is to be heated, or treated as a serious airlock. I look for at least one window in the door or the wall. Nobody needs a gloomy mudroom—mud season is dreary enough already.

One architect I know likes to station a mudroom on each of his buildings. It adds a beguiling touch to the structure and a personal stamp to his work. His hometown is peppered with his three-dimensional thresholds: tall and narrow, some with arched doors, some with curved roofs, some with steep gables. Most stand at homefronts, but one doubles as an airlock and clodbuster for a local tavern. Another of his mudrooms features a domed door with a tiny leaded window and a steep pitched roof. It looks like a sentry box at Buckingham Palace. No matter what form the mudroom takes, though, the end result is utility.

Although mudrooms may look a bit like outhouses, I will resist all temptation to carve a quarter-moon in mine. One embellishment I will add is shared by all the best mudrooms: the grate dirt remover. This is a salvaged heat register set into the floor in place of a door mat, preferably vintage cast iron to withstand the kicking and scraping of mud-caked boots. The register collects dirt. It can cover an inset cleanout box or an opening clean through the floor, although the latter may be drafty.

The bigger the mudroom, the more you can store in it. It's a good place to keep a cat or hang a side of beef (but not simultaneously). Some people install an unplugged refrigerator or an insulated storage box, to keep food cold, fuel free, in winter. (There's only one thing crazier than a hip roof, and that's artificial refrigeration during a Vermont winter. The mudroom gives you a reasonable alternative.)

Mudrooms intended for refrigeration keep coolest on the shadowy north side of the house. On the sunny side, a mudroom that's equipped with cold-frame glazing can yield extra vegetables in the spring and fall. Now there's a designer's dilemma with just one solution: the two-mudroom house. □

Tim Matson is a writer and photographer in Thetford Center, Vt. "Mudrooms" appears in Matson's A Country Planet, *published by Countryman Press, (Woodstock, Vt. 05091; $9.95).*

From *Fine Homebuilding* magazine (April 1986) 32:44-45

Photos: Tim Matson

Replacing an Oak Sill

Doing the job on a formal entry without tearing out jambs and trim

by Stephen Sewall

When I undertook the task of repairing the front entry of a Colonial Revival house in Portland, Maine, with its large 3½-ft. by 8-ft. door and side lights, I knew that the most difficult part would be replacing the rotted oak sill. It had suffered the neglect that many do, eventually checking and rotting from exposure to the weather because it hadn't been given periodic coats of sealant. The other repairs—which included replacing the raised panel in the door, replacing the pilaster bases, repairing the side lights and making some crown molding to replace part of the portico trim—were reasonably straightforward, but the sill presented some problems.

It seemed impractical to remove the entire jamb, replace the 7-ft. long sill and then reinstall it as the original (drawing, facing page) had been. Removing the jamb would require dismantling much of the trim inside and out, so it would be a big, labor-intensive job. Also, disturbing the entry that much could make refitting the door and side lights more difficult. Finally, I decided it would be best to replace the sill while leaving the jambs—both the inner jambs and the outer jambs at the side lights—intact.

Before the old sill could be removed, though, the door and the side lights had to be taken out. The side lights were held in place by stops on four sides. Since all of the trim was in good shape and I wanted to use it over, I was very careful to pry the stops out without damaging them.

The best tool I have found for removing any sort of wood trim is the Hyde #45600 Pry Bar-Nail Puller-Scraper. It has a thin blade that you can insert under almost any piece of trim without damaging either the molding or the surface to which it's attached. The thin end can also be sharpened so that you can cut small wire finish nails by hitting the other end of the bar with a hammer. The curved end of the pry bar can be used to open the joint up further and to scrape down the crusted paint before you reinstall the trim.

The stops of the side lights were inside mitered like most window trim, and because the side stops went in after the top and bottom, they had to be removed first. It's best to start prying at the middle, because the miters lock the ends in place. On each side light, I had to cut a few of the nails near the ends of the stop with the pry bar.

The author installed a new sill in this Colonial Revival entryway before going on to replace the raised panel in the door and repair the side lights and part of the portico trim.

As soon as the stops were removed, I marked their back sides so I'd be able to put them back where they belong. I pulled the remaining nails through from the back so I wouldn't disturb the finish side of the wood. The tools I have found most useful for this are a pair of end snips or end pliers. They shouldn't be too sharp or they will cut the nail instead of pulling it out.

After removing the side lights, I cut plywood panels to fit between the jambs. These would keep the jambs from floating free when I removed the sill. I made the panels short enough so that I'd have room to cut out the old sill underneath them (photo facing page). I attached the panels to the inner and outer jambs with drywall screws.

With the door and side lights removed and the jambs locked in place, I was ready to cut out the sill. I used a Sawzall with a 6-in. blade to make cuts through the sill on either side of both inner jambs, and as close to them as possible (drawing, facing page). I made two more cuts 3 in. from each end of the sill, being careful not to hit the nails coming from the outer jambs into the end grain of the sill. This let me remove three large chunks of sill, leaving only the four pieces directly under the jambs.

I split out the remaining sections piece by piece with a 2-in. chisel. With all of the wood removed, I was left with 20d nails protruding from the end of the jambs. To get rid of these, I used a metalcutting blade in the Sawzall where I could, and a hand-held hacksaw blade on the less accessible nails.

The old sill had rested on five equally spaced 1-in. strips of wood embedded in mortar and running the width of the sill. I removed them and the mortar so I would be able to slip the new sill under the tenons of the door jambs. I also chipped away the mortar line between the brick and the old sill so that new mortar could be worked in under the outside edge of the new sill once I shimmed it into place.

The new sill—I saved the best section of old sill as a pattern for the new one. After a clean, square cut on the radial-arm saw, I was able to trace a full-size end profile. But even if I could have found a 14½-in. wide piece of 10/4 stock, the replacement sill would have been too hard to make in one piece. The 2-in. raised section under the door meant ¼ in. of stock would have to be removed from the rest of the sill. Making the sill in two sections, one ¼ in. thicker than the other, and splining them together would reduce the work—and the waste.

With the widths of 10/4 oak stock I had available, I made up the sill out of a 6-in. wide piece and an 8½-in. wide piece. The 6-in. section became the part over which the door would sit. This meant that I had only about a 3-in. width over which I'd have to waste ¼ in. of stock. After face-jointing and planing the stock to a net thickness of 2¼ in., I rabbeted out the ¼-in. by 3-in. section on the jointer. (If you haven't got access to a jointer, you could do this on a 10-in. table saw with the blade fully extended.)

The flat, raised section under the door was beveled at 3° to the back of the stock on the table saw. I cut a 45° bevel from the flat under the door to the point at which the finish floor contacts the sill. I used a rabbet plane to cut the small bevel on the other side of the flat.

The 8½ in. wide section needed its front edge ripped at 93° from the face so that it would be plumb when the sill, which would slope slightly toward the outside, was installed. I routed the top front edge with a ball-

Stephen Sewall is an architectural woodworker in Portland, Maine. Photos by the author.

From *Fine Homebuilding* magazine (February 1984) 19:34-36

bearing rounding-over bit. A ⅛-in. by ⅛-in. drip kerf was cut under the front edge of this piece to keep water from finding its way under the sill.

With all of the bevels cut and the front rounded over, I made the cuts for the splines. I cut a ½-in. slot 1 in. deep on each section of the sill, with the top faces held against the fence of the table saw. I epoxied in a piece of Baltic birch for the spline.

The visible parts of the sill needed to be belt-sanded before the wood could be finished. It's worthwhile to scrape off excess epoxy squeeze-out while you're gluing up because it can dull sanding belts in a hurry once it's dry. I find that most putty knives spread the glue on the surface rather than pick it up. The flexible tip of a small artist's paint knife, available at most art-supply stores, works much better.

To finish the sill, I wanted to use something more durable than spar varnish with an ultraviolet filter. In checking our local marine supply, I found that I had a choice of two two-part polyurethane varnishes that are used on the topsides of boats—Petit Durathane (Petit Paint Co., Inc., Borough of Rockaway, N. J. 07866), and Interlux Polythane Super Gloss (International Paint Co., 2270 Morris Ave., Union, N. J. 07083). It is important to buy the thinner recommended for these products. The first coat needs to be thinned down, because the unthinned varnish is a thick syrup that won't penetrate sufficiently. Be sure to use these products in a well-ventilated area. They smell terrible, and the vapors aren't especially good for your health. I applied three coats.

Installing the new sill—The top of the pilaster bases were in the way of installing the sill, and they also needed to be replaced, so I removed them. The bottom of the pilaster itself was also in the way, but the distance between the pilasters was only ¾ in. less than the finish length of the sill. I used my Japanese *azebiki* saw (it has rip teeth on one side, crosscut teeth on the other) to cut small sections out of the pilasters. Its thin, flexible blade makes a clean cut with a narrow kerf. I set the pieces aside to be epoxied back on once the job was complete and the sill and pilaster bases were in place.

I cut the new sill to length with a skillsaw against a homemade guide clamped onto the sill. To fit around the outer jamb framing, I cut 2-in. by 8-in. notches out of the inside ends of the sill. These were cut partway with the skillsaw and then finished off with a handsaw. With these cuts made, the sill could be slid into place.

My new sill was a little thinner than the old one, and the old mortar had been cleared away, so there was room for the sill to slide in under the tenons of the door jambs. With the sill held up snug against the tenons, I marked the mortise locations on the sill. Before I pulled it back out, I took a rough measurement at the back edge to see how much it would have to come up to be at the correct height to the finish floor. This gave me a di-

Doorway with original sill

To start the job, Sewall removed the doorway's side lights and nailed plywood panels in place, with clearance beneath them to slide in the new sill. Then he removed the original sill, which had been resting on 1-in. strips of wood set in a bed of mortar.

Illustrations: Christopher Clapp

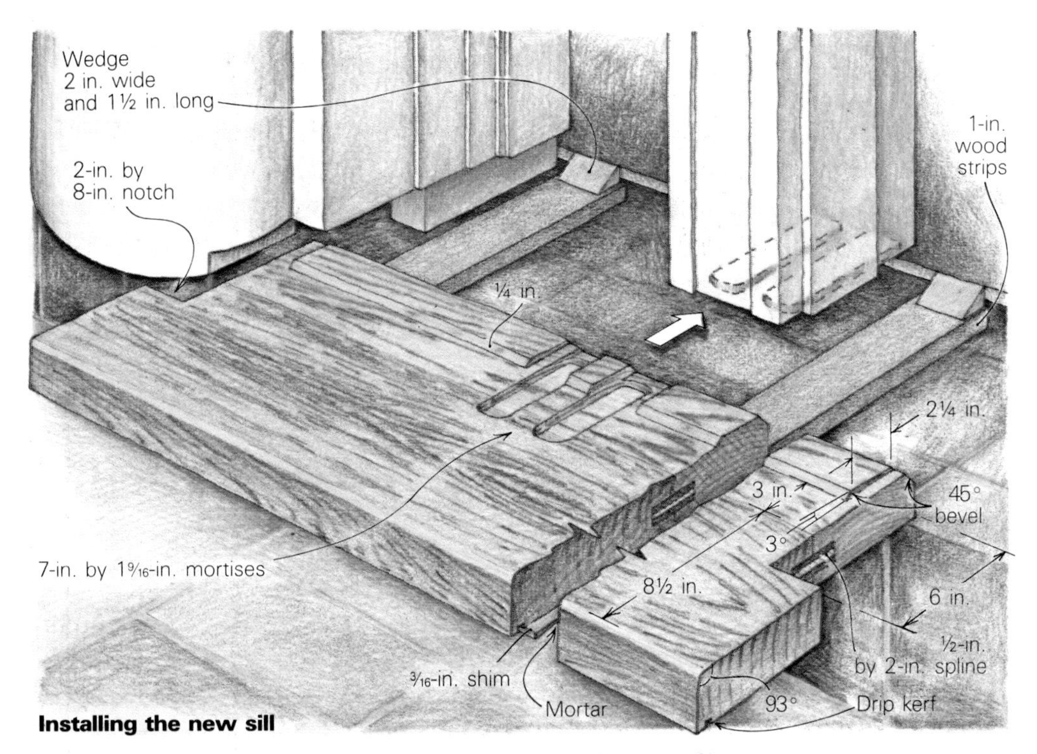

The new sill was splined together from two pieces of 10/4 oak, cut to the dimensions and angles shown in the drawing. The double mortises in the sill received the tenons at the bottom of the door jambs (photo left). Once the sill was in place, the pilaster bases were reinstalled (photo right), using shims to close up the joints between the middle and top half-round sections. The slivers that were cut out of the pilaster so that the sill could slide in were later glued back in place.

mension for the wedges needed to lift the back edge snug into place.

I used a router with a ½-in. straight-face bit to make the four mortises in the sill. I cut the mortises freehand, taking successively deeper cuts until the depth was just over ½ in. Each inner jamb had a double tenon. I cut each mortise ⅛ in. wider than the tenons and extended the mortises through the inside edge (drawing, and photo above left). The extra length was necessary because the wedges would start to push the mortise onto the tenon before the sill was all the way back into place. The extra length would be covered up when the inside plinth block was reinstalled.

I made two wedges out of oak to use for the back edge of the sill. They were 2 in. wide and 1½ in. long. I calculated one to be ⅝ in. high and the other ¾ in. These were set on the subfloor under where the sill would rest when it was slid into place. I used a hand plane to put a slight 45° chamfer under the back edge of the sill so it would slide up more easily on the wedges.

On the second dry fit, the sill slid up the wedges and I got a pretty good fit against the jambs, but one of the wedges needed to be changed to get a perfect fit. With the two wedges adjusted for height, I was ready to make three more to support the entire back edge of the sill. I ran a line from the top of the two wedges and took the measurements at the intermediate locations. With all five wedges in place, I tried a final dry fit. This time I used cedar shingles to shim up the front edge, where the mortar would eventually be packed. I needed to shim up the front edge only about ³⁄₁₆ in.—just right for a good mortar joint.

Before installing the sill, I spread epoxy on the tenons and in the mortises. I also ran a bead of butyl caulk underneath all of the shoulders of the jambs, and spread a stiff batch of mortar between the wedges and as high as I could without interfering with the sill's installation. As the sill rode up on the wedges and came tight to the jambs, the excess butyl and epoxy squeezed out and had to be cleaned off. I reinserted my four shims under the front edge, two at each end and two at the door jambs to lock the sill into place.

During all this, the door jambs were held in position sideways by the plywood in the sidelight openings, but they could move in and out slightly in the sill mortises. My next task was to make them plumb so that the front door would close properly. I ran a string on the inside from one outside jamb to the other, then tapped the door jambs into position with a block and hammer. To hold them there, I toe-screwed through them into the sill with 3-in. screws. I also put several toe-screws into the sill at the outside jambs because there was nothing else there to prevent the sill from creeping out, except the cedar shingles and the mortar under the front edge.

I mixed the mortar in a loose batch and worked it under the front edge of the sill. I snapped off the cedar shingles and knocked them in far enough so they would be covered by mortar. With a pointing tool, I worked as much mortar as possible back under the sill. I cleaned up the excess mortar with a stiff brush and water. The butyl caulk was scraped off, and the surface was wiped down with a rag soaked in paint thinner.

Finishing up—The old pilaster bases had been saved for samples so that new ones could be faceplate-turned on the lathe. I used mahogany because it is available in large dimensions and is more resistant to rot than pine. There were three pieces to each base: the square bottom, a half-round middle section and a scotia and half-round on top.

Before I installed the bases, I primed them with oil-base paint. To cushion the wood and to keep the square base from touching the granite below, I covered the base's bottom with butyl caulk and a layer of lead, which was fastened with copper nails.

The top section of the base fit into a rabbet at the bottom of the pilaster, so it could not be slid into place. I caulked the pilaster with butyl and held the upper section in place. With the square section put in place, the middle half round could be slid between the two.

To tighten up the three base sections, I inserted cedar shingles between the bottom two (photo above center). At some points on the circle no shimming was needed, and at others as much as ⅛ in. was needed. I used 8d galvanized finish nails to pin the sections to each other and the pilaster to the base. The cedar shingles were cut off as far in as possible with a utility knife, and the joint was caulked with butyl. Finally I epoxied and nailed back the slivers that had been cut out of the pilasters so the sill could be slipped between them.

I hung the front door as soon as the epoxy glue in the door jambs had hardened and the mortar under the sill had set. The door fit just as it had when I removed it. The side lights, stops and plinth blocks were reinstalled, and the job was done. □

Casing a Door

Work carefully, and save the wood filler for the nail holes

by Bob Syvanen

Nowhere in house building is the workmanship more obvious than in interior trim, and doors are among its most visible locations. It takes skill, patience and the proper tools to do good work. Perhaps more important, the use of some special techniques will lead to results that the craftsman can be proud of.

The wood most commonly used for trim is pine, though almost any wood will do—solid wood, that is. Various alternatives to solid wood are now available. Such materials as plastic and hardboard are more stable than solid wood and less likely to split. But while they may look like wood, they don't feel like wood, smell like wood or work like wood. There is really no substitute for the real thing. If you're going to work with the best, use #1 clear material.

Casing stock often comes in random lengths from 7 ft. to 20 ft. Usually, however, you'll find it in the standard lengths of 10 ft., 12 ft., 14 ft. and 16 ft. You can sometimes purchase complete trim packages for standard-sized windows and doors. A door package will include two head casings (the horizontal pieces above the door) and four jamb casings or side casings (the vertical pieces at the sides of the door).

Door casings come in various widths and profiles, but all are partially relieved, or plowed, on the back surface. (For a useful booklet on wood molding and casing patterns, contact the Western Wood Moulding and Millwork Producers Association, Box 25278, Portland, Ore. 97225.) The relieved section is about 1/16 in. deep, and leaves a shoulder about 3/8 in. wide on each side of this trough. When the trim is installed around the opening, one shoulder rests on the door frame and the other rests on the wall. The relieved section bridges any high spots that might be between them, and allows the casing to make good contact with each.

A quick way to relieve shop-made casings is to run the stock on edge through a table saw, with the blade set at a slight angle. If you turn the stock end for end and repeat the cut, a triangular section from the back side will be removed, leaving two flat shoulders like those on mill-cut casings. The drawing at right shows three styles of casing you can make in the shop.

If the head casing is to be square cut and laid on top of the side casings, it can't be relieved along its entire length because the relieved area will show up at the ends of the head casing. On such pieces, relieve the trim using a dado blade on your table saw. Start and stop the blade just short of the casing ends.

Tools—Any tool that makes a job easier will usually improve your work, since it will let you concentrate more on the result and less on working the tool. The best example of this I can think of is my miter box. Years ago, I used a homemade wood miter box. When I switched to a heavy-duty metal miter box with its own backsaw, I did better work faster. Now I use a power miter box. It speeds up the work and allows me even more time to concentrate on what I'm doing. (Power miter saws are sometimes called chop saws.) Now I wouldn't think of trimming out a house without one. The quality of a tool also makes a big difference. Good work cannot be done with a dull saw, a dull plane or a power tool that makes sloppy cuts.

In addition to the power miter box, the tools I use for casing a door are these: an orbital sander, a 13-oz. hammer, a block plane, a 2-in. and a 1/2-in. chisel, a small and medium nail set, a combination square, a framing square, a utility knife and a tape measure.

I also consider my portable work stool to be standard equipment for interior trim work. This stool, called a cricket, is simply a step made out of 3/4-in. material, with a shelf below to hold tools, nails, glue and sandpaper. Mine is about 14 in. wide by 25 in. long by 19 in. high, but the measurements can vary depending upon your needs. What's important is that you can stand on it and reach the head casing, and that you have convenient access to your tools. There should be a hand hole in the top so that the whole "casing shop" can easily be picked up and carried to the next opening.

Consider the opening—It's easier to case a door if the rough framing of the opening is square and plumb, with plenty of support for the trim. The thickness of the walls should be consistent, but often it isn't. The jambs may be poorly aligned or of different widths, and the wall finish may be sloppy—a real pain when you're trying to get casing to fit neatly to both jamb and wall. And all too often, sheetrockers will stop their sheets at either side of the rough opening and fill in the space above with a small piece. This usually causes problems. Most important for the finish carpenter is the thick build-up of joint compound at the corners of the door opening. Sometimes the excess material can be sanded down or scraped away, but it's usually difficult to case over.

A successful casing job also depends on how well the door jamb was installed. For casings to fit properly, jambs should be 1/8 in. wider than the finished wall thickness. This leaves the jamb 1/16 in. proud of the drywall on each side of the door, and allows you to bevel the edge of the jamb nearest the drywall (before the jamb is installed) so that the casing will fit snugly.

Casing the opening—The first step in casing the opening is marking the jambs for the setback, or reveal. The reveal is measured from the inside face of the door jamb (drawing, next page), and allows for a slight margin of error when installing the casing. I make the reveal about 1/8 in. to 1/4 in., depending on casing size and type, but it's mainly an aesthetic decision.

A combination square, set for the depth of the reveal and used as a gauge, can be used to mark the reveal. It works quite well, but I prefer to make up a wood gauge block. A gauge block won't lose its adjustment, as a combination square sometimes does. I make my block from a 2-in. by 4-in. piece of 3/4-in. thick scrap stock, with a rabbet cut in it on four sides to the depth of the reveal. To mark the reveal, I lap an edge of the block over the door frame, and run it along the frame to guide my pencil. The resulting line is quite accurate. The gauge block also makes it easy to mark the reveal at the corners

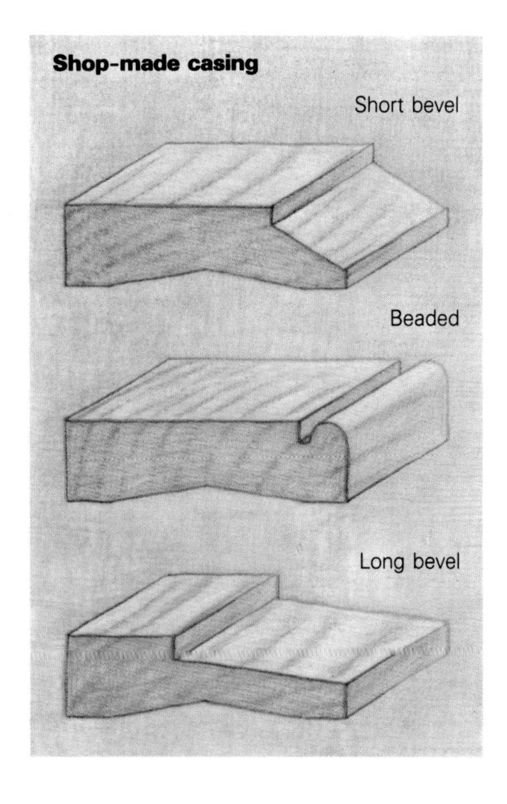

Drawings: Christopher Clapp

of the door jamb where a combination square can't quite reach. When two people are installing casings, it's a good idea to cut identical gauge blocks for each so that all reveals will be the same.

There are three basic styles of door casings: square-cut, mitered and corner block.

Square-cut casings. Simple square-cut casings (drawing A, facing page) are fairly easy to cut and install. Once the reveal has been marked, the casings can be cut to rough length and the pieces distributed to appropriate openings. I like to do this in order to be sure I have them all. It's upsetting to interrupt the flow of work in order to cut, shape or order one extra side casing. Working one opening at a time, I begin fitting the casings by making a square cut at the bottom of each side casing, checking each one for a precise fit at the floor after aligning it with the reveal line. When the casings must be installed before the finish floor is laid, I fit them to a scrap wood block as thick as the finish floor will be.

With one of the side casings held in place, I align it with the reveal line on the jamb, and mark it where it intersects the head reveal line. I use a sharp knife to make a precise mark on the inside edge of the casing, not on the face. A square trial cut can then be made about ⅛ in. above this mark, followed by a trial fit with the head casing. What's important here is to determine the correct angle. If the cut isn't quite right, an adjustment can be made when the piece is recut. The second cut is made a tad below the first one, but still shy of the true cut mark. When the angle of the trial cut is perfect, it can be duplicated at the true cut mark. The casing can then be aligned with the reveal line and tacked in place. Drive one 6d finish nail partway into the stud and one 8d finish nail partway into the jamb so that the casing can be removed for more trimming if necessary. Repeat the procedure for the remaining side casings.

The head casing is completed simply by squaring one end, holding it in place on top of the side casings, marking it to length with the knife, and cutting. Some people like to cut the head casing slightly long, so that it overhangs the jambs by ¼ in. or so.

After double-checking the alignment of the side casings, I nail them in place at the door jamb with 6d finish nails (4d for thin trim and some molded trim). Hardwood should be predrilled before you nail it. The casing edge that is against the wall should be nailed with 8d finish nails into the trimmer studs. I also like to toenail a 4d finish nail through the end of the head casing and into the top of the side casing. This helps to keep the joint surface aligned. The trim should fit tightly against the door jamb and the wall, so use as many nails as it takes to do the job. A good way to test the trim for tightness is to rap the casings with your knuckles. If they make a rattling "clack-clack" sound, put another nail in and rap again. When you hear a solid-sounding thump, the casing is tight. All nails should be set.

As an additional precaution against joint separation, I usually run a bead of white or yellow glue along wood-to-wood contact points just before nailing. These are very visible joints, and when glued and nailed, they'll hold together even if the wood shrinks. Wipe all traces of glue off the surface, using a damp cloth. This is very important if the casing is to be stained.

Variations of square-cut casing. One variation uses head casing that is thicker than the jamb casings. It looks best when these head casings extend past the jamb casings by the difference in their thickness (drawing B). For example, if the jamb casings are ¾ in. thick and the head casing is 1 in. thick, the head casing would extend past the jambs by about ¼ in. on each side. A second variation starts with a standard square-cut casing, and then adds mitered backband molding around the outside edge (drawing C). A third variation is used when the inside edge of a flat casing is beaded. This technique requires that you miter the beaded edge, square cut the ends of the head casing and cover the outside edge of the whole casing with a mitered backband molding (drawing D).

Mitered casings. The second, and most common, way to trim out a door is to miter the corners of the casing (drawing E). With mitered casings, a trial 45° cut is made instead of a square cut at the top of the side casings, but otherwise the trial cutting and fitting are the same as with square-cut casing. If the door jamb isn't square (use a framing square to check the corners where the head jamb meets the side jamb) and can't be adjusted, trial cuts on scrap stock will help you find the proper angle for the head casing. The side casing and head casing must have the same angle cuts in order for the outside corners to meet and for the molding profiles to match. This step is very important if the casing will be stained, since the joint is so visible.

When the outside corner of a trial miter joint is open, the angle of the next trial cut will have to be adjusted to take more stock off the heel, or inside corner, of the miter. Remember that head and side casings must be cut using the same angle. The adjusted cut on the head casing, therefore, should reduce the gap by one half, with the remaining half cut from the jamb casing. These adjustments are usually so small that changing the saw angle with any precision is difficult.

To make fine adjustment cuts using a miter box, I put a wedge between the miter-box fence and the stock—pieces of cardboard, wood chips or even plane shavings will work. By moving the wedge away from or closer to the saw blade, slight variations in the angle of the cut can be made. Another way to make fine adjustments is to use a block plane. Make sure the iron is sharp and set for a fine cut.

Mitered casings are installed in much the same way as square-cut casings. The only additional nailing that's required is at the outer portions of the mitered corners of the head and side casing, where 4d finish nails or brads should be sunk from each direction.

Corner-block casings. The third method of trimming out a door is to fit the ends of head and jamb casings to corner blocks (drawing F). The corner block should be a little thicker than the side casing, and can be combined with a similarly shaped plinth block. The plinth block's purpose, aside from making the visual transition between the casing and the baseboard, is to provide a wide, flat surface for the baseboard to die into.

The plinth block is installed first, and should be plumb. It can be installed with a reveal that matches the casing reveal—⅛ in. to ¼ in. is fine. Installing the casing from this point is the same as described earlier, except that the side casings are fitted and butted to the plinth block.

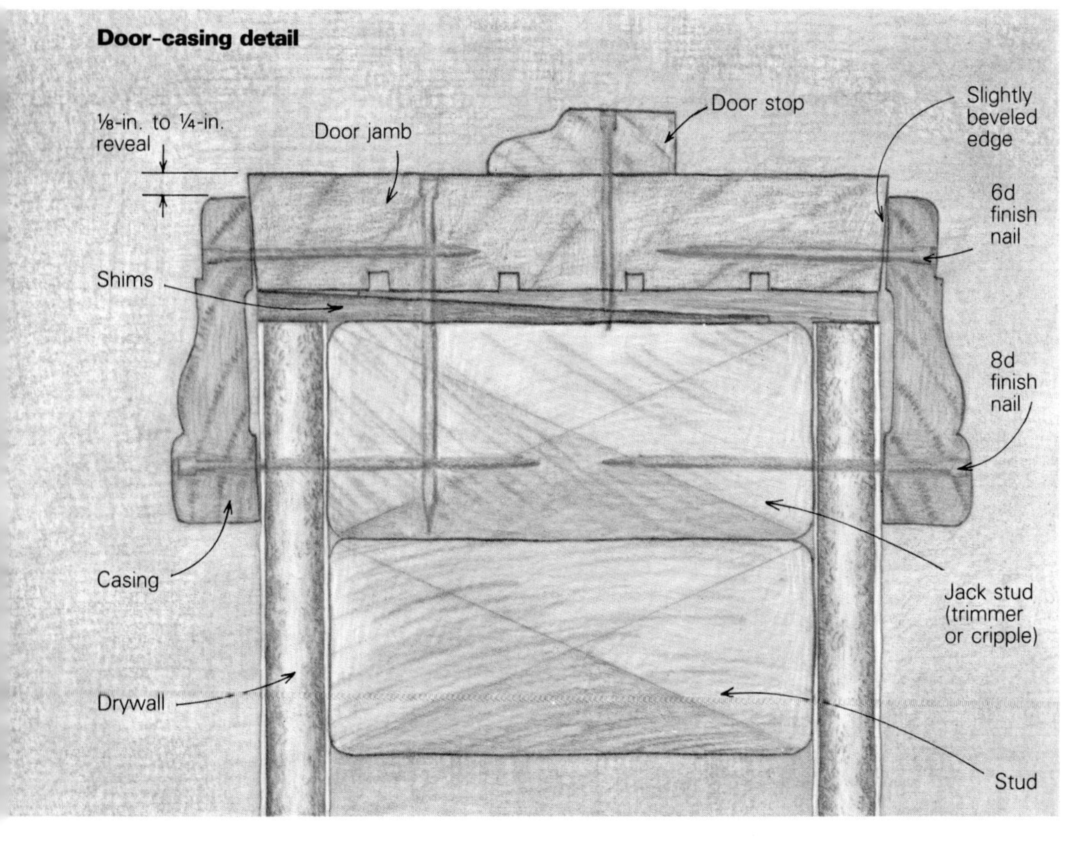

From *Fine Homebuilding* magazine (December 1985) 30:55-57

Mating the surfaces—These techniques all work well when the walls are flat and evenly dimensioned around the door opening. But since a house isn't usually built to the tolerances of furniture, wall surfaces won't be perfectly smooth, and the openings won't be textbook square and plumb. Fortunately, there are ways to compensate for irregularities.

With square-cut casings, a little tipping away from the plane of the wall can be tolerated, but not much. When the top of a head casing tips toward the wall, the entire length of the joint in front will be open. To correct this, put a wedge-shaped shim behind the head casing to bring it into proper alignment. A spot of glue will hold the wedge in place.

You can also shim mitered casings, but when the tipping is minor, the back side of the 45° angle can be shaved with the block plane to bring the trim into alignment. This sort of adjusting cut can also be made in a miter box by placing a shim between the casing and the bed of the miter box, near the blade. The resulting back cut will allow the front joint of the miter to close up. If the front edge of the miter must be trimmed down, a wedge placed under the casing at a point farther from the blade will do the trick. All these cuts are made long for a trial fit before cutting to exact length.

When the thickness of the wall is greater than the width of the jamb, the casing will tip in toward the opening, resulting in a visible gap between the casing and the wall. The surfaces of the casing will be out of alignment as well. The gap can be eliminated by planing the shoulder of the casing to fit the wall surface. The alignment problem can be a bit more difficult to deal with, but not impossible. Unfinished casings with flat surfaces can be surface planed with a sharp block plane and sanded with an orbital sander or a sanding block. If the trim is to be stained, however, great care must be taken to avoid sanding across the grain, as the marks will show in the finish. Molded casings can be trimmed to align at corners using sharp chisels.

Wherever shims are used or a casing is planed, it's important to adjust the other trim pieces that will be affected. For example, if a side casing is thinned at the top, the head casing must be thinned to match. If the side casing is thinned at the bottom, then the adjoining baseboard must be thinned, as well.

Filling nail holes—I find that painted casings require more effort to prepare for than stained casings. Paint accentuates any imperfections, and slight dimples left by partially filled nail holes show like beacons through the paint. So the filling must be done carefully. For tools, you'll need a putty knife, good eyes and a sensitive hand.

The best filler I've found was recommended to me by a painter. ONETIME Spackling (Red Devil, Inc., 2400 Vauxhall Rd., Union, N. J. 07083) is an acrylic spackle, and its smooth, putty-like consistency makes for easy application. It dries quickly and sands easily, so you can paint over it almost right away. Using the putty knife or a dab on your finger, fill the holes as completely as you can (if you've done the casing correctly, you won't have any joints to fill). After the spackle dries (a couple of minutes), use fine sandpaper to smooth away the excess. Check the result by rubbing the filled areas with your hand—you'll feel imperfections you can't see. Spackle again where necessary, sand and check. If the first coat of paint reveals a nail dimple, reach for the spackle again.

When the job calls for the casing to be stained, filling nail holes goes a bit differently. I find it's easier to stain the casing before I install it since it's easier to do the job at my workbench. After the casing is installed and the nails are set, I brush on a coat of polyurethane. This protects the wood, and will keep the putty I use from staining the wood around the nail hole. Once the polyurethane has dried, I fill the nail holes with Color Putty (Color Putty Co., 1008 30th St., Monroe, Wis. 53566). It comes in many colors to match various stains, and you can also mix colors to match unusual stains. I rub it into the hole with a putty knife or the back of my thumbnail, and remove the excess with a rag or my finger. After the putty has dried, I apply at least one more coat of polyurethane to the casing.

Refinements—Before finishing, all corners and edges of the casing should be eased. Sharp edges don't hold paint well, and they dent easily and are uncomfortable to touch. I like to block-plane these edges before installing the casing, using one continuous stroke for each cut. A continuous stroke leaves a smooth surface that keeps the same bevel along its length. If you have to plane an installed casing, you'll need a bullnose plane, a chisel or sandpaper to get at the inside corners.

Occasionally a piece of casing won't look as good as you think it should, so you have to remove it. If the piece has to be used again, the nails should be pulled through from the backside of the wood. This is particularly important when working with prestained trim. If you drive the nail out head first, it will tear the surface of the wood as it exits. To pull nails out from the backside, use either a hammer with sharp edges on the claw or pincer-type pliers. The hole can easily be filled later.

When I'm through casing a door, I load my tools on the cricket shelf and move the whole thing to the next opening. I never leave an opening until it's completely finished and ready for the painter—it's too easy to forget something otherwise. □

Bob Syvanen, of Brewster, Mass., is a consulting editor of Fine Homebuilding *magazine.*

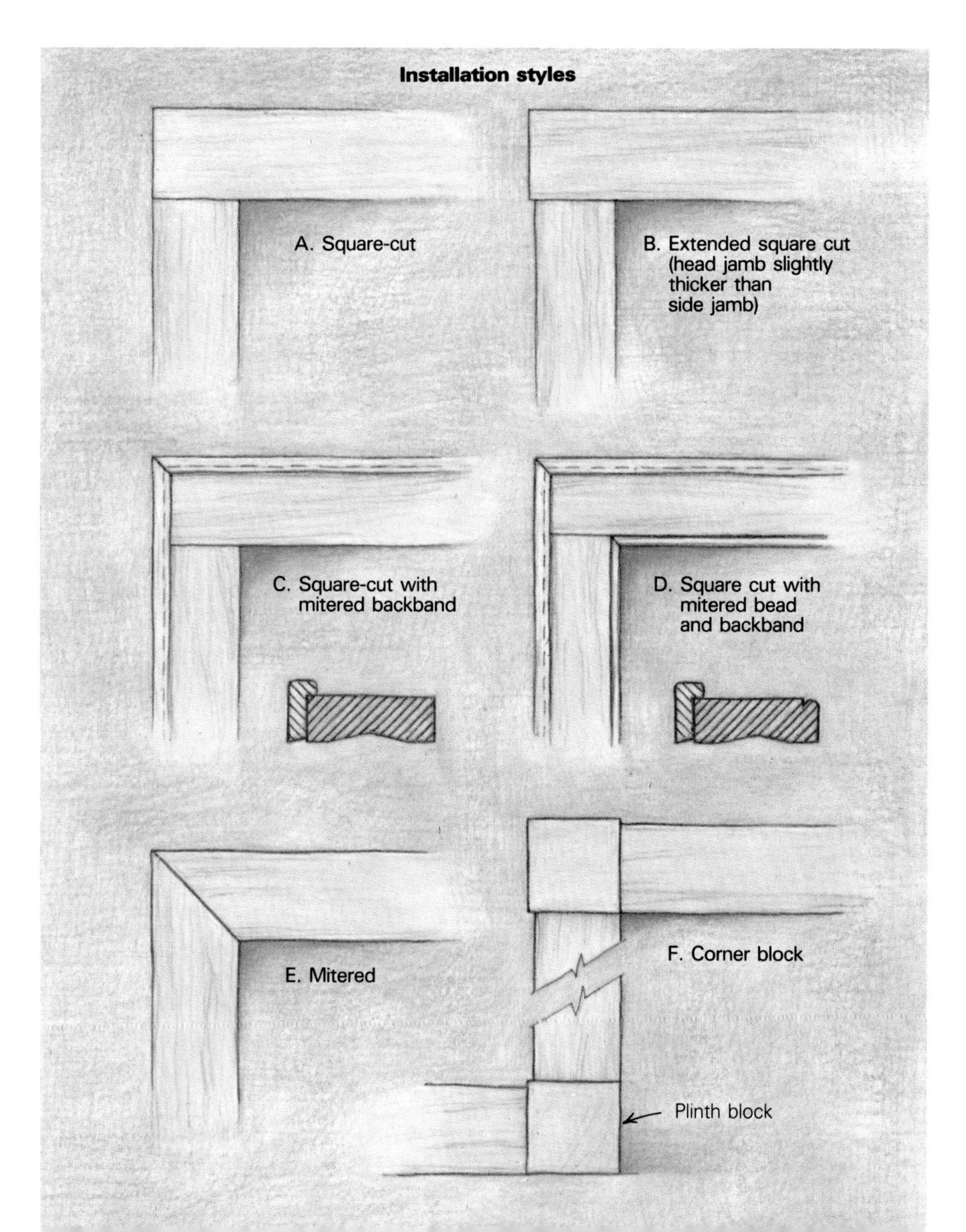

Stripping Trim

Why and how to take old paint off interior woodwork

by Mark Alvarez

You don't always have to strip moldings, baseboards and trim. Sometimes, the best approach is to clean and paint over what's already there or—in extreme cases—to rip everything off the wall and install new stuff. But stripping trim is a job all renovators face at one time or another. It's never neat or easy, but there are efficient and effective ways to approach it, and the results can be spectacular. Your house may be a Victorian, for example, with fine hardwood trim that was meant to be seen in its natural state but has been painted and repainted over the years. It's a prime candidate for stripping. Even if you have simple pine or fir trim that was meant to be painted, it may be peeling and cracking or have so many layers of paint globbed over it that its details have been obscured, and its visual effect ruined. If the wood's still in good shape, it makes sense to strip it down to a clean, smooth surface before you repaint.

There are three basic ways to remove paint from wood: abrasives, heat and chemicals. Abrading paint off with sandpaper or those spinning wire wheels you chuck up in your electric drill is either very slow (orbital and hand sanding), very hard to control (belt sanding), or both (wire wheels). It's tough to get into crevices and cracks with these tools, too, and they all kick up a lot of paint dust. If your house is 30 years old or more, this dust will probably contain lead, a hazardous material you don't want to breathe. So always wear a dust mask if you decide to sand.

The heat plate that is so commonly used on house exteriors works only on flat surfaces. It won't do the job on moldings, in corners or along the hard-to-reach edges of your trim. Heat guns—those giant electric hair driers that generate astonishingly high temperatures—are better for trim, but on curved surfaces especially, inexperienced operators almost invariably end up scorching some wood. The heat these machines put out can crack glass, too, so you shouldn't use them around windows. With any heat method, there's always some danger of fire. Never use a propane torch or any other open flame.

Chemical paint removers—strippers—are either caustic or solvent. Caustic strippers, which are usually used in commercial dip-stripping operations, are lye-based—usually sodium-hydroxide or potassium-hydroxide formulations—and the residue they leave on

the wood must be neutralized with an acid after the finish has been removed. They are sold either as powders to be mixed with water, or as pastes.

Caustic strippers are cheap, but they darken wood and raise its grain, so you shouldn't use them if you intend to leave your trim clear. They are often used by professionals when the trim is going to be repainted, but many clients specify that the contractor run pH tests to be sure of adequate neutralization. Many restorationists feel that caustic strippers can never be satisfactorily neutralized, and that they will eventually begin to bleed out of the wood and cause paint failure.

Most experts agree that solvent removers, which have a methylene-chloride base, are best for stripping interior trim, especially if you've got a lot to do. They are more expensive than caustic strippers, and they present hazards of their own, which will be discussed below, but they do a thorough job without harming the wood.

Tom Schmuecker's family owns Bix Process Systems, Inc., which manufactures stripping products for the general public and also trains and licenses dealers who use the products professionally. Schmuecker strips wood professionally himself, and over the last few years, he's seen his business, once almost entirely centered on stripping and refinishing furniture, expand strongly into renovation and restoration work. He's stripped a lot of trim, and has developed techniques that keep the job straightforward and manageable.

Getting ready—Solvent removers aren't as dangerous as lye, but you should always protect your eyes with goggles. Schmuecker also wears rubber gloves, coveralls and old shoes. And he makes sure that pets are kept outside.

The fumes of solvent strippers can make you dizzy and sick, so you need good ventilation when you work. A couple of open windows are sometimes enough, but you might want to set up a fan. Strippers work best at room temperature, but it was cold the day we took these photos, and while Schmuecker had a couple of windows open, he didn't want to turn on the fan and lower the room's temperature too much, so he wore a respirator with an organic-solvent filter. If you do the same, remember that the room should be thoroughly ventilated once you're done. Don't heat the area directly. This is a good way to make the remover dry out quickly, limiting its effectiveness. Schmuecker also had a bucket of cold water and a mop handy in case of a spill. On larger jobs he relies on a wet/dry vacuum.

It's easier to strip wood in the workshop than on the walls, so on big jobs Schmuecker likes to take down as much of the trim as possible—crown molding almost always, because it's so hard to get at—and haul the pieces to his own shop, where he can strip it in a tray. Then he returns it to the house and nails it back up. This is an especially good idea if other parts of the room have already been restored or refinished, because it keeps the mess in the shop. But before you do it, consider the damage you may do to the surrounding wall, and how easily it can be patched.

Schmuecker has found that if he's careful, he can pry trim off old plaster walls without harming them. But if you've got original plaster walls in perfect condition, you might decide to work on the trim in place. To strip the window shown here, he removed the window guides and scratch-marked them (stripper removes ink and crayon) so they could be returned to their original places. He stripped these pieces and the lower sash in his tray. On other jobs, he's removed every foot of molding and baseboard.

Schmuecker advises planning remodeling or restoration work so that you can proceed from the top of the room down. Scheduling subcontractors is always a problem, though, and he's had to strip trim in a number of rooms whose floors had already been expensively refinished. Even if your floor has not recently been returned to glory, you'll want to protect it from stripper and paint residue. Schmuecker first lays down a few layers of

From *Fine Homebuilding* magazine (August 1983) 16:66-69

newspaper, then covers them with plastic sheeting, firmly duct-taped around its edges, snug up to the baseboard. As an added precaution, he lays down a painter's dropcloth over the other layers. The newspaper acts as a final line of defense in case the dropcloth and plastic are punctured, and also keeps the plastic from slipping. The wall near the trim also has to be protected. Schmuecker uses plastic here, too, taping it down first with masking tape and then with duct tape. He also covers radiators and the like with plastic sheeting (photo right).

Testing—There are different kinds of solvent strippers. Some are flammable and some are not. Some require a solvent or a water rinse, and some require no rinse at all. For most of his work, Schmuecker prefers non-flammable removers that need to be rinsed. Flammable removers are cheaper, but not quite as effective on paint, because they contain between 30% and 50% methylene chloride as opposed to 50% to 85% in the non-flammable paint removers. No-rinse strippers are fine for overhead work, for example, where rinsing liquids would drip on you and the floor, but they just don't work as well as rinse-removers against either paints or applied stains.

Once he's set up, Schmuecker always does a test patch, applying remover to a small portion of the wood to be stripped. If the molding is intricate, he tests its most ornate section, which is usually the most troublesome area. The patch verifies that he's using the right stripper, and gives him some idea of how many layers of paint he's dealing with, and how long it will take to remove them all.

At this point, if the wood is undistinguished, its location is awkward, and the finish promises to be tough to remove, Schmuecker sometimes recommends simply replacing it. Anything can be stripped, he says, but there are times when it just isn't worth the trouble.

Some paint removers work faster than others. Strippers that raise quick blisters are usually breaking bonds between layers of paint, or between the first coat of paint and the wood surface. These fast strippers work fine if there are only a few coats of paint, especially if they were applied over a varnish that sealed the wood. Slow-acting removers, on the other hand, dissolve paint rather than breaking bonds. If left on long enough, they can go through six to eight layers at a time, and they also get paint out of unsealed wood better.

If he's working in several rooms, Schmuecker does a test patch in each, because the woodwork in different rooms will probably have different tales to tell. Kitchen trim, he says, has usually been repainted many times and will probably require the use of a slow-acting remover, but bedroom trim will probably have only a few coats of paint on it, and a faster stripper might be the best choice. This information is especially important to a pro, for whom time is money.

Professionals know which strippers will harm which materials. If you aren't sure, and

Applying the stripper. **Schmuecker wears goggles, rubber gloves and coveralls when he works with paint remover. The room isn't well ventilated, so he's wearing a respirator, too. Plastic sheeting protects the walls. He's run a test patch at the junction of sill and casing. Each application consists of two coats of remover; a light one followed two to five minutes later by a heavy one.**

In areas where the paint remover might drip off the surface, it can be covered with cellophane. A cellophane cover also lengthens the stripper's wet life and keeps it working longer.

After the paint remover has been applied, it can be worked into corners and gaps with a knife blade to get as much paint as possible out of these tough spots.

Before starting to scrape, Schmuecker makes sure that the paint has been dissolved down to the bare wood. The stripper he uses may be renewed with another light application if it dries out.

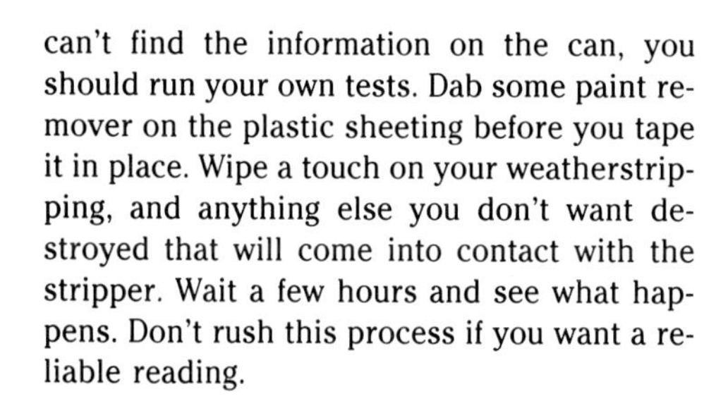

can't find the information on the can, you should run your own tests. Dab some paint remover on the plastic sheeting before you tape it in place. Wipe a touch on your weatherstripping, and anything else you don't want destroyed that will come into contact with the stripper. Wait a few hours and see what happens. Don't rush this process if you want a reliable reading.

Applying the remover—On a small job, Schmuecker simply pours stripper into a coffee can and works with a 2-in. to 3-in. brush. For heavy paint, he applies a light coat first, letting it stand for two to five minutes, and then covers it with a much heavier coat. He brushes the remover on in one direction only, so as not to thin it out and encourage evaporation. To lay on the thickest possible coat, he dabs, rather than sweeps, with the brush.

On larger jobs, he often sprays the stripper on. Sprayers are professional tools, and cost anywhere from about $200 to about $800. Don't try to use a common paint sprayer. It's important that the tank, gaskets and hoses be compatible with the remover you're using. You shouldn't, for example, use an aluminum tank with a methylene-chloride stripper. The metal will react with the chemical, and the tank could burst. If you have questions about the compatibility of a sprayer and a stripper, write to the paint remover's manufacturer, and request its material safety data sheet.

Whether he's brushed or sprayed the remover on, Schmuecker covers it with Saran Wrap to hold it against the surface in spots like the underside of the window rails, where the stripper might drip off (photo top left). The plastic wrap also retards evaporative drying and increases the stripper's wet life by up to an hour, so Schmuecker covers his remover whenever he's working on a small area with an especially thick paint layer.

After 10 or 15 minutes, he uses a knife or an ice pick to work remover into cracks, corners and the most deeply carved parts of ornate trim (photo above left). Then he lightly recoats the area.

It may take as long as four hours for the surface to be ready for scraping. Many people begin to scrape the remover off much too soon, before it has had a chance to penetrate and dissolve the layers of paint. By doing this they work a lot harder than they have to, and use much more stripper than necessary. Schmuecker usually coats more windows or trim while he's waiting, but he cautions that it's very easy to get ahead of yourself. Scraping takes much longer than applying remover, so by the time you've come back and scraped a few windows or feet of trim clean, the last areas you coated could have dried out. Work only a window or a few feet of molding ahead until you can judge your speed.

The instructions on some strippers tell you not to let them dry out. This is because they'll harden if they're allowed to dry, and then you've got to start all over again. Others can be reactivated with another light layer of remover. The strippers Schmuecker works with can be reactivated.

Scrape, rinse and clean—It's time to scrape off the remover with its paint residue when a test scrape shows you're either down to bare wood (photo above right) or that the lowest layer of stripper is drying out. This is the most tedious and time-consuming part of the job. It invariably takes a lot longer than an inexperienced person thinks it will. The first time you strip molding, give yourself twice as much time as you think you'll need. Begin by gently scraping the flat areas with a putty knife (photo top right). Drop the sludge into a pan or bucket. Schmuecker uses a disposable aluminum baking pan, which is big, light and cheap. In corners, a soft brass brush (photo center right) or a stiff plastic brush works well. Old dental tools are best for the most ornate areas. They come in all sorts of odd and helpful shapes, and they're best for stripping just at the point when they're no longer good enough to be ground down again for use by your dentist. A shavehook (photo center left) is especially good for getting paint out of dents and dings in the wood.

After you've scraped and brushed, there will still be a few stubborn spots of paint left. If you are going to repaint, you can stop here and apply the rinse, but if you're after a clear surface, recoat and scrub the wood with a plastic steel-wool substitute (the real thing can leave particles embedded in the wood to oxidize, change color and leave a pebbled finish that looks like black-eyed peas).

Tool catalogs offer scrapers shaped to fit some standard moldings. It's worth buying one if it fits the trim you're working on. When he has to strip hundreds of feet of a single, unusual molding, Schmuecker takes a section of it to a machine shop and has a scraper ground to fit the pattern. He says the frustration and time this saves is well worth the $30 to $40 for the custom grinding.

Sometimes, if the paint is very thick, another complete application of remover is necessary. Just repeat the whole procedure, using both the light and heavy applications. The five or six coats on the window shown here came off with a single application, and Schmuecker proceeded to rinse. You can use either water or solvent—lacquer thinner or paint thinner—to rinse the wood down. Schmuecker prefers solvent, even though it's flammable. It dries much more quickly, and it doesn't raise the grain or make the wood swell. He rubs on a light coat with the steel-wool substitute, then wipes the surface down with a damp cloth and lets it dry (photo bottom right). The trim is ready to refinish in a day, after a light sanding with 220-grit paper.

As anyone who's done any stripping knows, it's next to impossible to get paint out of joints. This is no problem if you're going to repaint, but is trouble if you want to leave the finish natural. Schmuecker doesn't even try to remove all the paint. Instead, he scrapes it back, much like a mason raking a mortar joint, and fills the joints in either with putty, which he later stains, or with colored lacquer sticks or wax crayons.

Another common problem is finding heat or water marks, mottled wood or stain residue on wood that you want to leave clear. Schmuecker recommends applying—right after the final rinse, if possible—an oxalic acid solution (½ cup to 1 gal. warm water). Let it stand for two or three minutes, then rub it down with lacquer thinner, paint thinner or water. This will lighten the marks, mottles and residue so that the wood can take a stain. □

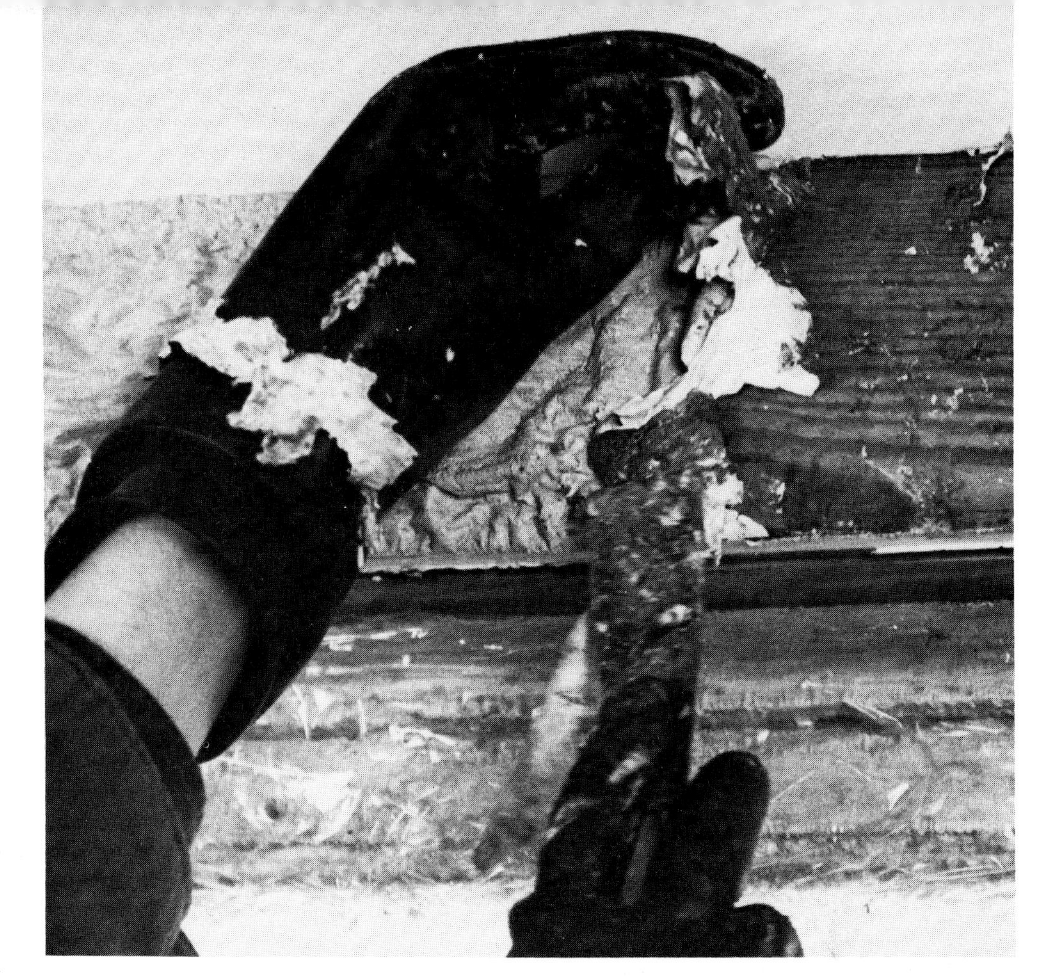

Paint-stripping tools. **It sometimes takes as long as four hours for a heavy coating of remover to dissolve paint down to the bare wood. Once it's time to scrape, a putty knife (top) will lift globs of paint and stripper off the surface. A shavehook (above left) and a plastic or brass brush (above right) come in handy for corners and edges.**

A good scrub with synthetic steel wool gets off the last specks of paint. If you're going to repaint, you don't need to do this. The last step is to wipe the surface down with solvent, as above.

Spray-Painting Trim

Preparations and priming make or break the job

by Byron Papa

Painting is often considered a trade that doesn't require much skill. I guess that's because almost anyone can pick up a brush and apply the paint, whatever the outcome. It may not loom large in a construction budget, but the painter's work almost always shows up more than anyone else's. When you walk into a house and look around, most of the visible surfaces are finished with some type of a coating that has been applied by a painter. A truly professional job can upgrade the quality of a house, and even be the deciding factor in a sale.

Every painter has opinions as to how best to apply paint, stain and varnish, and indeed, there are many ways to get the job done. Some of my methods are pretty ordinary—some are not. In this article, I'll talk about painting in general, some of the tools and equipment involved and how I finish trim with oil-base enamel paint.

There is a spraying revolution going on in the painting business today. Many suppliers nowadays even have spray-equipment departments. Most of this hoopla stems from the development and maturation of the airless paint sprayer, which can handle thick house paints and coatings better than past contenders. Some painters spray nearly all of their work; diehards never get near the gadgets. In my own work, I strike a balance. I certainly use spray equipment, but for residential jobs I find its usefulness to be somewhat limited, especially for exterior work. Most of the time when I do spray, such as with doors and trim, it's for the quality of the finish rather than the speed of the tool, since the extra masking and cleaning up usually tip the scale against any time saved.

On most new houses, the walls are painted first, followed by trim. I do just the opposite, and here's why: Common flat latex wall paint is easily damaged during construction. If walls are done early, carpenters, plumbers, electricians and all the other crews will have more time to smudge and gouge them. This can be frustrating and costly to the painter, and can lead to confrontations with other subs.

Second, I paint trim with a spray gun, and this way I don't have to worry about getting overspray on the walls. Trim benefits from spraying more than the walls do because most trim paints and varnishes have a glossy finish, which will magnify brush marks. Also, paintbrushes con-

---

Byron Papa is a home builder from southern Louisiana. He is currently building in Huntsville, Ala.

stantly pick up dust and spread it along the work. Using a spray gun, I can paint a whole house full of doors and trim in a few hours, and all of the preparation can be done alongside other crews.

The major drawback to doing trim first is having to go back later and mask it off to paint the walls, but since I paint the walls with brushes and rollers, I can get a clean edge with a single strip of tape.

Health hazards—There are health hazards associated with painting, and I do my best to protect myself and my co-workers from them. For sanding and dusting, I use 3M #8710 disposable non-toxic particle dust masks. These can last a week or more, and they block dust particles better than the other masks I've used. They do cost more than the budget versions and they can be hard to find; try the 3M consumer information number for local distributors—(800) 328-1667. I have had problems with their rubber straps coming loose after only a couple of days of use. I remedy this by smearing a dab of white glue around the area where the straps connect to the paper filter.

When I'm using a spray gun, I wear a twin-cartridge Binks respirator (Binks, 9201 W. Belmont Ave., Franklin Park, Ill. 60131). Its charcoal filters work for nearly all kinds of paint and varnish, but they have to be changed fairly often. Generally, when I begin smelling paint through the respirator, it's time for a change. I also wear a cheap knit hat when I spray, to keep paint mist out of my hair.

Different woods—Painters don't often get to choose what gets painted and what gets stained, but the type of wood being used for each job can make a big different in the results. I build my own houses now, and whenever I have the choice, I finish my trim with enamel paint and save the varnish for cabinets and bookcases. Most door casings and trim moldings are made from less expensive softwoods, such as white pine. These woods take on stain poorly. If the clients are set against painted trim, I'll at least try to persuade them to go natural—varnish on unstained wood.

For the best enamel-paint job, you want a tight, closed-grain wood. White pine works pretty well, and is widely available. I think Douglas fir is lousy for painted trim—its grain tends to rise and it requires a lot of sanding. My favorite trim wood is parana pine. Since it grows near the equator (mainly in Brazil), it is almost without annual growth rings. Its smooth texture takes enamel paint exceptionally well. If the cabinets are to be enameled, birch and basswood are good choices.

General preparation—With my system of painting trim, preparation is about 90% of the work. I start by taking down all the doors and removing their hinges and hardware with a cordless drill and the appropriate bit. Next I number the doors with an ink marker. I write the number on the doorknob mortise or on top of the door (not on the bottom, because some doors may be cut shorter for carpets before they are rehung). Then I mark the strike mortise on the jamb with the corresponding number. I also label hinge mortises on doors and jambs with a #2 or #3, designating the number of pin loops on the hinge plate. This is not absolutely necessary to get them back together, but it does make it easier. For bifolds, I label each pair with the same letters, and I draw arrows on their top rails that show where they meet.

I don't want the marks to be obscured by the paint, so I cover each one with a piece of masking tape. Then I gather all the hardware and put it in small buckets, keeping various screw sizes separated, and cover the buckets with aluminum foil to keep paint overspray out. I store the buckets out of the way in a fireplace or tub.

I lay the doors across sawhorses to remove their hardware, and while they're in that position, I drive a pair of 12d common nails into their top and bottom (drawing, below), about 6 in. in from the edges. The nails act as handles, and they bear on the sawhorse crossbars. This allows me to paint one side of the door, and then rotate it to paint the other side without having to wait for the paint to dry.

Sanding—Sprayed enamel builds up a fairly heavy coat, which is usually enough to hide slight surface imperfections such as planer marks. So I don't ordinarily sand trim before I prime it. If I'm going to apply a stain or varnish to a piece of trim, I'll sand out the planer marks with 150-grit paper.

For solid, raised-panel doors (usually white pine or fir), I use a one-third sheet orbital sander for the flat surfaces and sand the rest by hand. If they're to be painted, I use 100-grit, but if a natural finish is desired, I go no coarser than 150-grit. The same goes for flush veneer doors. Sometimes the edges of doors are a little rough, so I carefully belt-sand them with 80-grit or 100-grit. I also smooth over the corners of each door edge with a piece of 100-grit.

Hardboard door prep—Hardboard doors, such as the ones made by Masonite, usually come factory-primed. But I don't always find the primer coat to be adequate. It's often thin and spotty, and I've seen them with unprimed edges. Sometimes I just spot-prime these doors, but if a primer job is bad enough, I reprime the whole door. Sanding the factory coat isn't usually necessary. A good dusting will do.

A nagging problem I've had with hardboard doors is that often the corners are damaged and the substrate fibers are exposed and frayed. Unlike wood doors, a light sanding doesn't put them back in shape. And if they aren't smooth, the edges of a hardboard door won't seal well enough to receive the enamel top coat. Instead of sanding, I use a small-radius roundover bit to clean up their edges. The operation doesn't take very long, and I do all the doors, whether they're damaged or not. The rounded edge is

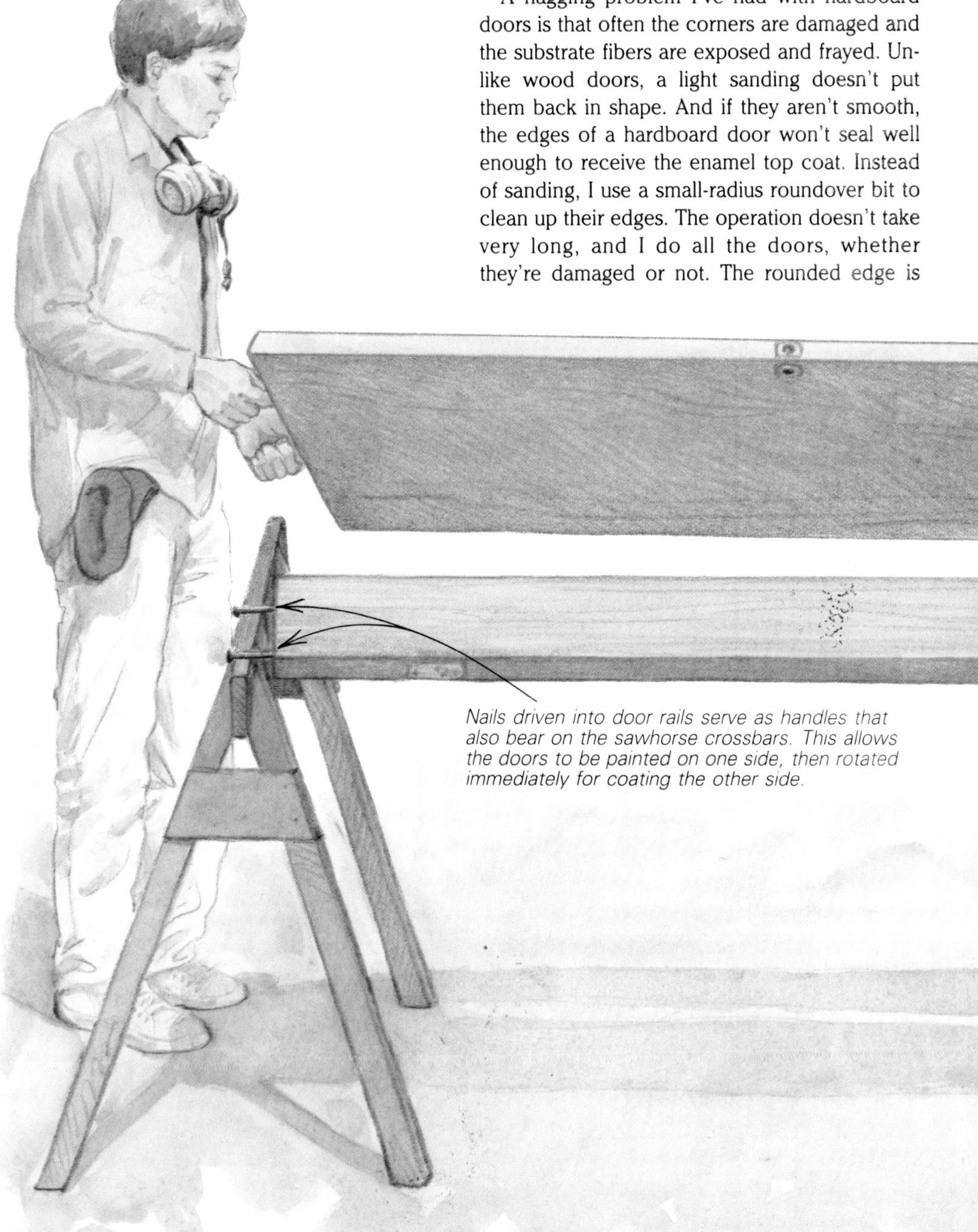

Nails driven into door rails serve as handles that also bear on the sawhorse crossbars. This allows the doors to be painted on one side, then rotated immediately for coating the other side.

Drawings: Jackie Rogers

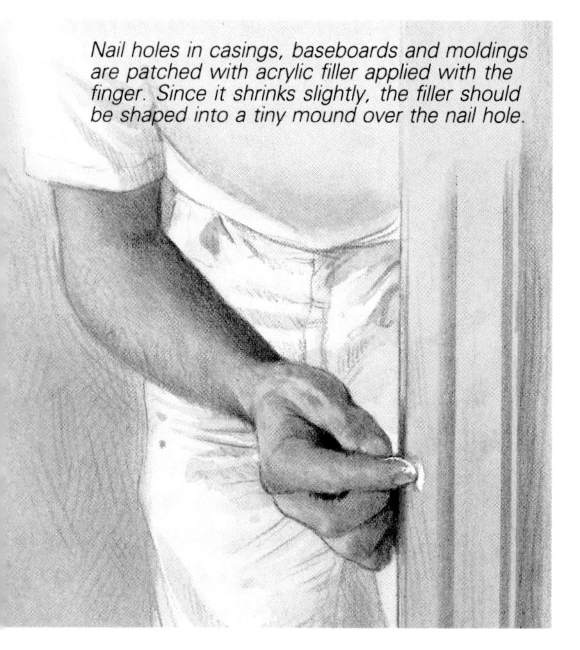
Nail holes in casings, baseboards and moldings are patched with acrylic filler applied with the finger. Since it shrinks slightly, the filler should be shaped into a tiny mound over the nail hole.

less susceptible to future damage, and it takes a coat of paint better than a sharp corner.

If you do this, be careful when approaching the hinge and latch-bolt mortises. If the bit's pilot slips into them, the cutters will gouge the door edge. I usually skip these areas and finish up later with an 80-grit sanding block.

Filling nail holes—When the trim is to be painted, I fill the nail holes first. There are two common types of premixed fillers on the market. While they are both latex based, the new "lightweight" variety uses tiny air-filled glass balls as aggregate. The lightweights dry faster and shrink less than the old formulas, but they are softer and shouldn't be used to patch vulnerable areas.

A putty knife can be used for filling big dings like hammer dents on a jamb, but for nail holes the filler is most easily applied with a finger, as shown in the drawing above. I keep a wet towel on hand to clean the excess off my fingers. After I push the filler deep into the hole, I smear it flush. Next I build up a small mound of filler over and around the hole to allow for shrinkage. When it's dry, I sand lightweight filler flush with 150-grit or even 220-grit paper. For the old style I use 100 grit.

Sometimes sanding will reveal small pits on the surface. These are trapped air pockets. To avoid them, apply the filler from side to side when you make the little mound, instead of pushing it directly on top of the hole.

Masking—One of the most time-consuming tasks that precedes spraying is masking the areas not to be painted. Often this simply amounts to protecting a couple of fiberglass showers and some window glazing. But sometimes the effort it takes to protect the unpainted surfaces overshadows the benefits of a spray job. An extreme case would be a room with natural roughsawn wood on all the walls next to a painted plate rail. Not only is that a lot of area to mask, but tape doesn't stick well to rough wood. Whether to spray in cases like this just depends on how much trouble the painter is willing to go through. For a nice cabinet, it might be worth it. I once erected a plastic "tent" to spray a built-in china cabinet in an existing house. It worked out well.

For all masking chores, it's important to use good-quality tape. A good tape should stick to the work, even when it comes in contact with strong paint solvents. It should also be easy to remove, even weeks later. Lately I've been using Sherwin-Williams' Professional Quality tape. It is one of the best brands I've ever found. Automotive paint supply centers usually have very good tape, but the price can be high. For most work, I like to use 1½-in. tape—it's a good universal size. Remember that even a good tape can be difficult to remove if it's been exposed to direct sunlight for more than a few days.

Besides tape, I keep on hand a horde of single-edge razor blades. They're cheap and they cut clean. For protecting large objects like tubs and showers, I use inexpensive lightweight plastic dropcloths taped along all edges. A 4-mil thick 9-ft. by 12-ft. dropcloth costs about $.75. For large glass and other flat areas, I use newsprint paper. I buy the end rolls from our local newspaper for next to nothing. The ones I get are about 2½ ft. wide and a single roll can be enough for several houses.

Masking window and door glass usually takes the most time. If the glass areas are big and don't have too many muntins, I use paper, taped down securely at the edges (drawing, below). If I'm masking a French door with scads of lites, I use tape exclusively. I use two strips of 3-in. tape, followed by a strip of whatever is needed to fill the remaining space. I start at the top and run each strip of tape all the way down the door or window at once, tucking it tightly against the muntins with my pocketknife as I go. After all the tape is applied, I come back with a razor and cut the tape at each point where it meets the muntins and peel away the unwanted pieces.

Window lites are masked with newsprint bordered by masking tape, which is allowed to run long, and then trimmed with a single-edge razor. Papa runs a penknife along the intersection of the tape and the mullions to make sure the tape adheres to the glass.

It can be aggravating for a painter to have to stop in the middle of spraying a coat of primer to mask off something somebody forgot, so when I think everything is taped, I go into each room and take another good look around, imagining the sprayer at work.

When I'm done masking and sanding the filler, it's time to dust and vacuum. An air compressor with a blower attachment works great for dusting door casings and doors, but I usually just use an old paintbrush. After cleaning all the work to be painted, I vacuum the floor out to about 1 ft. away from all the baseboards and door casings. I also use a crevice attachment to suck up any dust that may be trapped under the baseboards and casings.

Paint sprayers—Conventional air spray systems that work with air compressors aren't used very often in house painting. They're mainly designed for spraying thin coatings, and they are easy to use. I use my Binks Model 7 to spray cabinets and bookcases with lacquer, varnish and occasionally, very thin enamel paint. It's one of the most reliable tools I've ever owned.

Airless sprayers are far more popular for houses. They can handle heavier coatings, less thinner is required, they're faster and they don't require an outside air compressor. While conventional sprayers use compressed air to atomize the paint, airless sprayers simply pump the paint at high pressure through a hardened steel tip that atomizes it into a usable pattern.

The most common airless sprayers are hand-held cup guns. Most of them use a tiny piston pump that is activated by a small electromagnetic motor, which makes a loud buzzing sound. I started out using these little sprayers (Wagner 350s) over a decade ago, and they worked out okay, but they do have limitations. They're pretty slow and can be finicky at times. I learned early on to discard the "cone-spray" tips that are usually supplied with the guns and replace them with the longer-lasting tungsten carbide "fan-spray" tips that spray a flat pattern. A cone pattern is more difficult to control than a fan pattern, yet ironically, most of these small cup guns are intended for first-time users.

I still use airless cup guns for small jobs, because they're so easy to prepare and clean, but for most of my spraying I use an airless pump system. It consists of a heavy-duty stationary pump (drawing, facing page) tied to a lightweight gun by a length of high-pressure hose. This is the serious painter's tool. They are expensive (in the $1,500 range), but they can often be rented at contractor's rental yards.

Manufacturers use various designs for their pumps, and each has claims of superiority. For a small painting business like mine, I prefer the diaphragm pump. It costs less initially and is

From *Fine Homebuilding* magazine (October 1987) 42:54-59

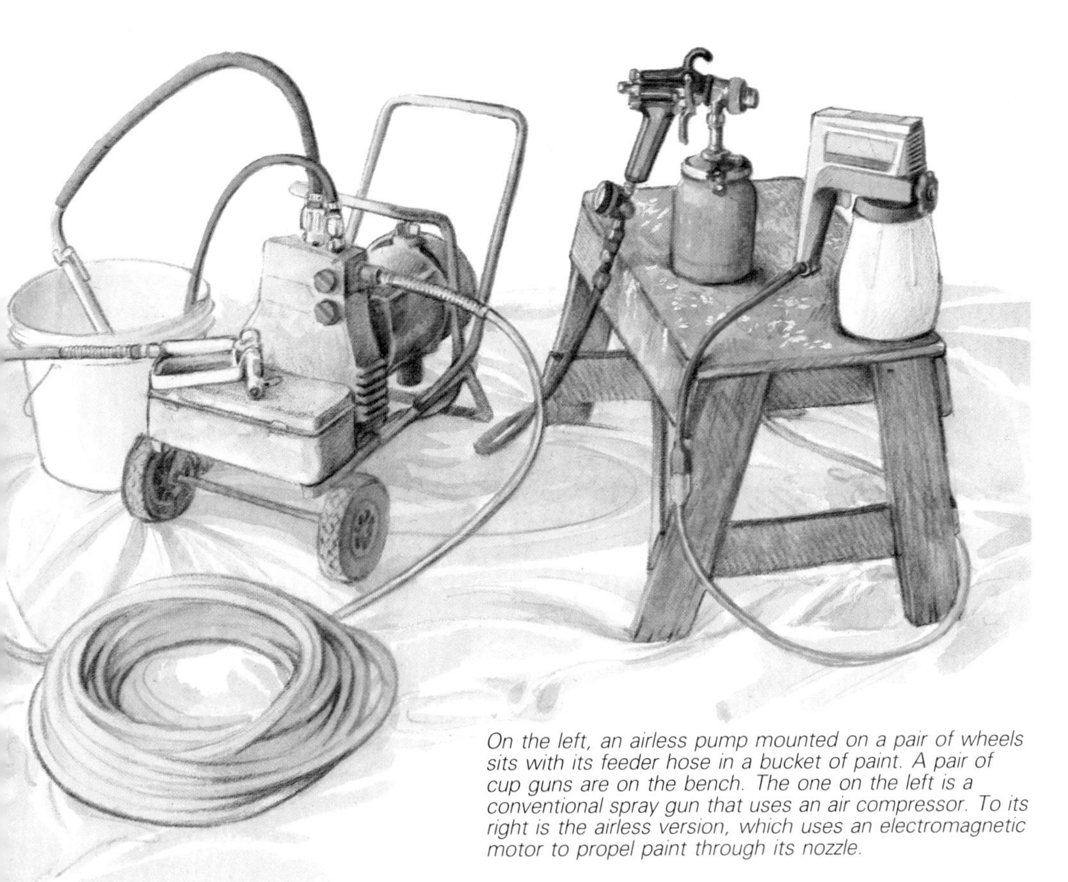

On the left, an airless pump mounted on a pair of wheels sits with its feeder hose in a bucket of paint. A pair of cup guns are on the bench. The one on the left is a conventional spray gun that uses an air compressor. To its right is the airless version, which uses an electromagnetic motor to propel paint through its nozzle.

easy to repair, even on the job. I have an old Wagner/Spray Tech 1000 Super that I bought in 1977, and it's still kicking. I stock most of the parts that are likely to fail and carry them in my paint truck. I've replaced just about everything at least once and don't plan to junk it until the motor gives out.

But airless sprayers do have disadvantages. First and foremost, they can be very dangerous. Because of the high pressure involved (2,500 psi or more), paint can actually be injected through the skin, causing serious injury (with a possible risk of amputation). Guns and tips are usually equipped with a safety tip guard, but this doesn't completely remove the danger. Users often remove the guards (me included) because mist can accumulate on them, causing drips. It's a good idea to keep your local poison-control center phone number on hand, just in case. The national number is (412) 681-6669.

Airless pump systems generally spray at a fast rate and can be difficult to control, especially in tight areas. Also, their tips wear out fast and they are expensive to replace ($15 to $30 apiece). A worn tip will spray unevenly and will tend to use more paint. A nagging problem is tip clogging—the slightest solid particle can block the paint flow. Manufacturers have dealt with this problem in several ways. My favorite solution is the reversible tip. When a particle gets lodged, you just flip the tip around, pull the trigger and it blows out. Flip the tip back around and you're ready to spray again.

Besides styles, there are many different sizes of tips available, and you have to use the right size for each type of paint or varnish. The orifice can vary in size from .009 in. to .050 in. or more, but for most house painting .011 in. to .017 in. is the common range. I use 0.11 for varnish and lacquer, .013 in. for interior wood primer and enamel, and .015 in. or .017 in. for heavier stuff like exterior primer and latex paint.

Also, each orifice size is available in a variety of fan-spray patterns. This refers to the actual width of the fan-shaped stream of paint measured in inches, at a distance of 1 ft. from the tip. I use 6-in. to 8-in. fans for most of my work—that's a good universal range.

Though both airless and conventional cup guns are easy to clean, cleaning an airless pump system takes time and a lot of solvent, so the job size has to justify its use. I usually don't fire up the airless for anything less than three gallons.

Another spray tool I often use is the aerosol spray can. It's hard to beat for small primer, varnish and lacquer jobs and touchups. I don't mind using them because nowadays many manufacturers offer the new FANSPRAY heads that actually spray a true fan pattern, just like a professional gun. On most of them the tiny valve nozzle can be rotated to change the fan from a vertical plane to a horizontal plane. These heads are truly amazing and in my view, make the old ones obsolete. They've been around for a few years, yet their presence is more the exception than the rule.

Before you spray anything—When I first began painting houses commercially, I learned a lot about the trade from reading. Half of what you need to know is written on the back of the paint can. Most labels describe the product, give specific preparation and application instructions, and warn of potential problems and dangers. A lot of mistakes can be avoided by checking this information before starting work.

Most paint manufacturers make available other literature about their products and how best to apply them. Sometimes this information is available in the store for the asking; other times you have to write for it (ask for a specifications manual). I use *Consumer Reports* magazine for information too. It usually has at least one or two articles a year on some type of paint or related product with brand-to-brand comparisons.

Primer—For most interior priming, I use an ordinary alkyd (synthetic oil-base) enamel undercoat. Every paint company makes one. I've tried several brands and they're pretty much alike, though you can sand some of them a little sooner than others. Lately, I've been using Sherwin-Williams Wall and Wood Primer. It sands well the next day. I use a fast-drying oil primer, called Kilz (Masterchem Industries, P.O. Box 2666, St. Louis, Mo., 63116) for special odd jobs when I'm in a hurry. But for the most part I find it gives off too many fumes for the bulk of my work, and it doesn't sand as well as the old-fashioned alkyds.

New paint is usually pretty clean and can be used as is, but older stuff might have to be strained through a filter. Inexpensive paper funnel-shaped filters can be used for small amounts. For larger quantities, nylon sack-like strainers are usually available from paint suppliers in both 1-gal. and 5-gal. sizes. I've also found that a large kitchen strainer is good for removing dried lumps and other big particles. It can be cleaned and used over.

The day before I spray, I mix all the primer I'll need for a job. In estimating, I always figure a little strong—running out to the store for more paint in the middle of spraying a job can be very disruptive. I mix the primer in 5-gal. drywall buckets and thin it about 15%, making it a little thicker than heavy cream. It's better to have a little too much thinner than not enough. Without enough thinner, the paint film won't level out properly. Too much thinner can cause the paint to run, but this can be overcome by spraying lighter coats. Ordinary paint thinner can be used, but I prefer naphtha. It evaporates at a much faster rate, which helps in two ways: the paint film sets up faster, and has less chance to sag or run.

Spraying primer—I spray two coats of primer on wood trim. This leaves a heavy enough film to smooth out slight imperfections in the texture of the wood, and enough primer after sanding to provide a good seal.

To begin, I set up the pump near the house entrance, and start spraying at a point farthest from it (usually the bedroom areas). The hose is 100 ft. long, so I can usually reach any place in the house from this setup. Airless paint hose is very stiff and can be difficult to handle without running it into freshly painted work, so it's best to have a helper assist in guiding it around the house as you spray. I also use a 30-in. "whip end" hose next to the gun—it's more flexible.

I spray a medium to light coat first, just enough to cover the wood. The adjustment is in how long a pass the gun takes over the work. The first coat is more likely to run than the second coat, so it can't be applied too heavily. While it's setting up, I have someone go around with a paintbrush and light to check for sags and runs. They'll most likely occur at overlap points, like the mitered corners of casings (drawings, below). Ventilation helps the paint set up more quickly, so I always crack a few windows.

When the first coat is set (preferably dry to the touch), I spray on the second coat, which is considerably heavier. Some of the solvent from it dissolves into the first coat, and reduces the chance of sags and runs.

To keep everything moving, I do all the doors and any loose moldings between the first and second coats on the trim. I always try to do them outside, using the sun and fresh air to promote faster drying.

After I've sprayed one face of the doors, helpers turn them over immediately, handling them by the nails. This way, I never have to stop spraying. I spray two coats on wood doors, just like the trim. I leave the doors on the horses, lying flat, until the primer has set up. This eliminates sags and runs. If the sun is shining, I won't let them stay out long because they might warp.

A note of caution: When spraying outdoors, always watch where the wind takes the overspray, so as not to damage nearby property. Though the airless-sprayer industry from its inception has been boasting "less overspray," the overspray mist that is produced is made up of larger particles than that of conventional sprayers, and it shows up more on cars and the neighbors' houses.

Sanding the primer—I sand all the primed surfaces with wet/dry silicon-carbide sandpaper. It holds up well, has a stiff backing and isn't affected by humidity. It's more expensive, but it lasts longer than cheaper papers. To save money, I buy standard 9-in. by 11-in. sheets by mail order, for about half of what they cost in paint stores, from Industrial Abrasives Co. (642 North Eighth St., Reading, Pa. 19603). The grits I use most often are 220 and 320.

Primed wood is a little rougher than primed hardboard, so I use 220-grit for sanding it. That's about as coarse as you can go without the risk of scratches showing through the enamel top coat. To get the most out of a sheet, I fold and cut it with a razor into four strips 2¾ in. by 9 in. I fold each strip into thirds, which makes a stiff, easy-to-handle pad. When folding sandpaper, never let a grit side contact another grit side. Otherwise, as they rub together, much of the cutting edge is destroyed before it's even used. Shallow planer marks and other minor defects will usually sand out with a few passes.

For the primer on hardboard and steel doors, 320-grit is usually coarse enough. Heavier grits can go through the thin primer. I use an aerosol spray can of fast-drying Kilz primer to touch up any areas I sand through.

Factory-primed steel doors are usually smooth enough as is, but I never trust their slick surfaces to anchor my new paint. So I scuff them thoroughly with 320-grit to ensure a good bond. Just before spraying, I wipe them down with naphtha to clean and degrease their faces.

Caulking—After sanding, I fill cracks between molding joints with latex caulk. On painted trim, even the slightest crack can stand out as a dark line, especially on light colors. The nozzle of the caulk should be cut small and as clean as possible. I use a razor, and shape the end to a point. This way, it will leave only a tiny bead.

I do only one crack at a time because latex caulk sets up quickly. I moisten my finger with a wet towel, and then I use it to smear the fresh bead of caulk smooth. This has to be done right away or the caulk will roll into tiny solid par-

Spraying a casing

The size tip I normally use for spraying oil primer and enamel is usually designated as #1308, which denotes an orifice .013 in. and a fan-spray pattern width of 8 in. at 1-ft. distance. I adjust the fan to spray vertical. To spray a typical casing, I step to one side of the doorway and stand facing it. I start at the top (drawing, below left), holding the gun perpendicular to the face of the trim molding and about 1 ft. away from it. Trigger pulled, the gun must always be in motion.

The hardest thing about doing a casing is the top corners—you're not able to start beyond the corner, a common rule of spraying paint. If you did start beyond the corner for the header and the vertical casing trim, there would be an inevitable overlap with too much paint build-up. Therefore, the skill lies in pulling the trigger and moving the gun at the same time. Some overlap still occurs, but it's minimal.

I start at the top left corner and move the gun to the right. The trigger should be released upon arriving at the other corner. Next I spray (from right to left) at

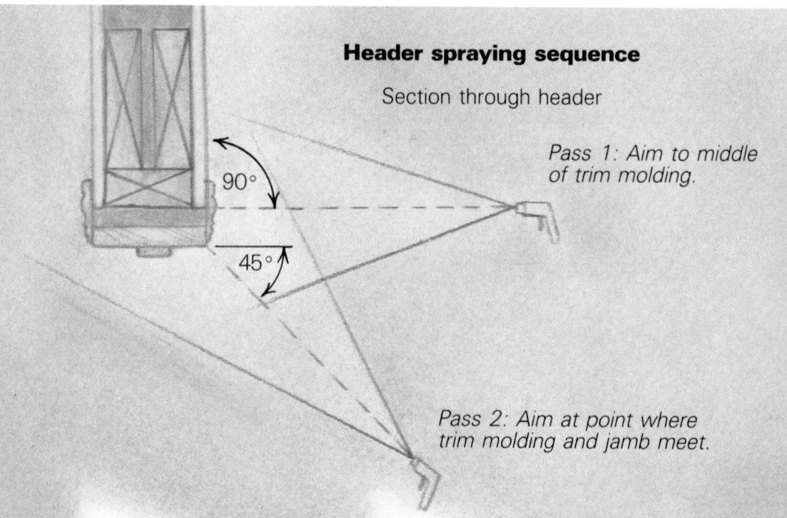

Header spraying sequence

ticles. If the caulking job is done well, the moldings will look like one, with no evidence that caulk was even used.

Applying the enamel paint—I prefer to use a semi-gloss oil-base enamel for my top coat. Oil paint sprays and levels out better than latex paint. It is also more durable, easier to clean and it has a richer luster.

I thin the enamel 10% to 15% with naphtha, depending on its consistency. I mix it all at once, as with the primer, but I try to estimate the amount required more accurately. Before spraying, all the surfaces to be painted should be dusted with an old paintbrush and the floors should be vacuumed again.

I spray enamel in much the same way as I do primer, applying the first coat somewhat lighter than the second. The second can be sprayed as soon as the first becomes tacky. If it looks too rough, it may need more thinner. Adding more thinner can partially compensate for a worn spray tip.

I follow the same procedure for the doors, but I usually allow exterior doors to bask in the sun longer, so they can be handled and rehung by the day's end. Popular wisdom warns against painting in direct sunlight, but I've never had paint blister because of this, though in some climates it may be a problem.

When all the painting is finished, I close up the house to keep the dust down. Unlike the primer coat, the top coat can't be sanded out. The enamel is usually dry by the next morning, but is fragile and can be easily damaged for a few days. I remove all the masking tape a day or so after spraying.

Brushing paint—If you use a brush, its quality is critical. I use Praeger white China bristles (model #W-104) for oil paints. They are the best bristle brushes that I have ever come across, at any price. I use a 2-in. brush for most enamel trim, but I also keep 2½-in., 3-in. and 4-in. brushes on hand for other chores.

Brushing enamel takes a lot of time, and though it's hard to cover the work with just one coat, we always try. Most good-quality paints can cover in one coat if they're applied to an ideal flat surface, but on a job this is rarely the case. I find that the worst problem with doing trim, a door casing for example, is that the bristles edge around corners and slide sideways onto freshly coated work. This almost always creates streaks. To help avoid streaking, I stagger the parts of the casing as I go. For example, in the morning I might do the outside edge of the casing trim molding, skip the face of the same molding, then do the inside of the molding along with the reveal, skip the jamb and do the edge of the door stop. I wait until the first sections are dry (even until the next day) before filling in. Doing it this way, you must wipe errant brush strokes off areas not being painted at the time.

Cleaning up—To clean up oil paint, I use a low grade of mineral spirits. I get it from a petroleum-products distributor for about $1.50 a gallon. GoJo hand cleaner works well on the skin, and for the unavoidable overspray that gets on around my eyes, I use petroleum jelly. Sometimes I rub it on my hands and face before the spraying begins, to aid cleaning later, but not if I'm going to be handling some of the work to be painted—oily hands threaten adhesion.

I get first-class results from spraying my trim this way, but it's a dirty job, and I'm usually pretty glad when each "spray day" is over. □

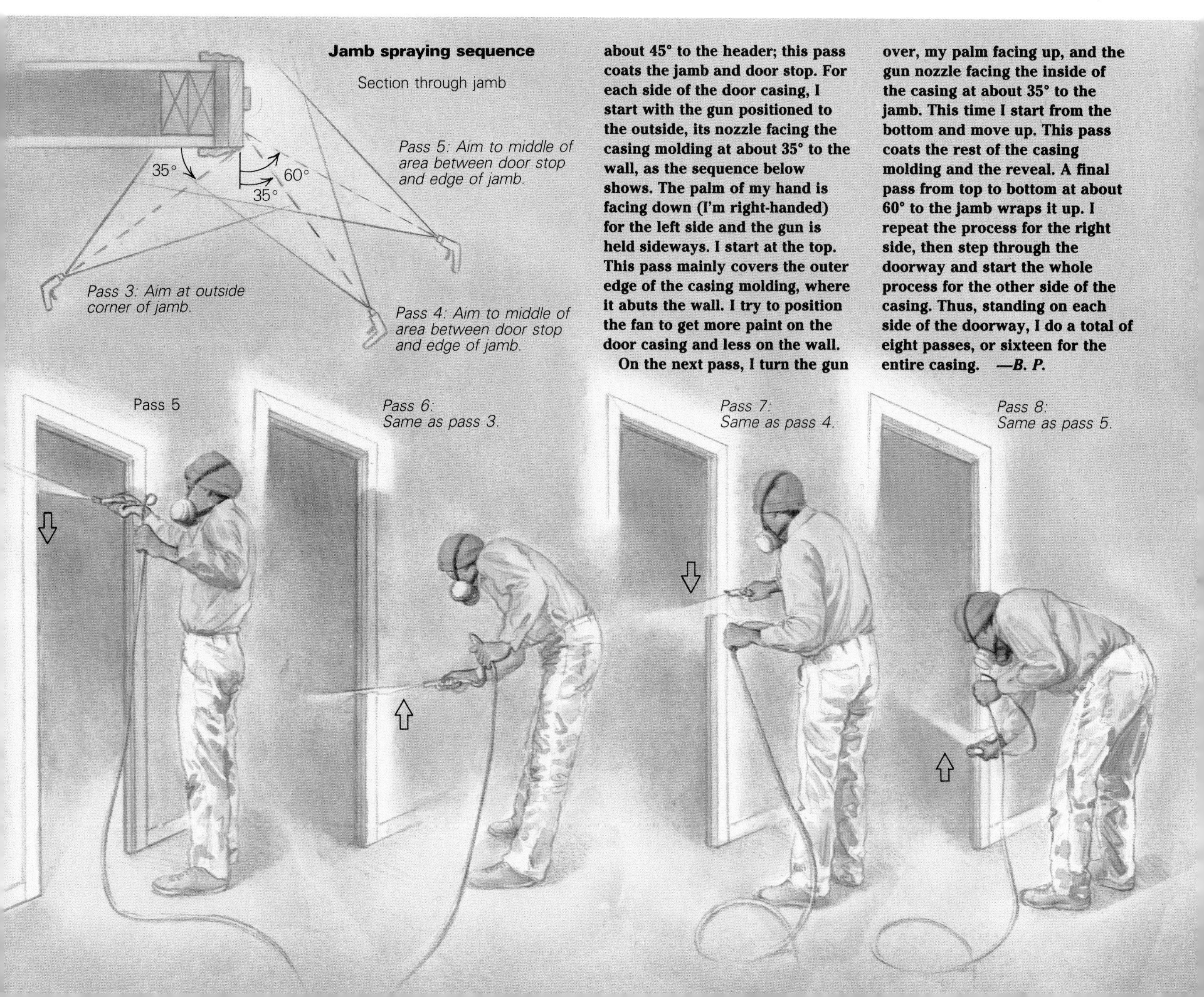

Jamb spraying sequence

about 45° to the header; this pass coats the jamb and door stop. For each side of the door casing, I start with the gun positioned to the outside, its nozzle facing the casing molding at about 35° to the wall, as the sequence below shows. The palm of my hand is facing down (I'm right-handed) for the left side and the gun is held sideways. I start at the top. This pass mainly covers the outer edge of the casing molding, where it abuts the wall. I try to position the fan to get more paint on the door casing and less on the wall.

On the next pass, I turn the gun over, my palm facing up, and the gun nozzle facing the inside of the casing at about 35° to the jamb. This time I start from the bottom and move up. This pass coats the rest of the casing molding and the reveal. A final pass from top to bottom at about 60° to the jamb wraps it up. I repeat the process for the right side, then step through the doorway and start the whole process for the other side of the casing. Thus, standing on each side of the doorway, I do a total of eight passes, or sixteen for the entire casing. —*B. P.*

Making Window Sash

How to do a custom job with ordinary shop tools and a router

by John Leeke

On a historic-restoration project I worked on not long ago, the house's window sash were in poor shape. The original plan for the sash was to repair the worst of them and then replace them all sometime in the future. But before the window work started, the owner decided to have new sash right off. I didn't have enough time to place an order with a custom millworks; so I decided to make the sash in my own small shop, even though it lacks specialized sashmaking machinery. This meant I'd have to match the joinery and molding profile of the originals, and I'd have to work quickly enough to make money on the job without overcharging my customer.

The old sash were hand-made over 150 years ago. One had been without paint for many years, so its joints came apart easily. All I had to do was see how it was made, and reproduce each part. The challenge was to keep track of all those parts, and to make the joints fit properly so the sash would get as much rigidity from its mechanical integrity as from the glue in the joints.

It takes me about 5¾ hours for all the setups needed during a run of sash. The production time for the kind of sash described here on a short run of three or four double units is almost six hours per unit. Considering my time and the cost of materials, the final price was about 20% higher than ordering custom-made sash from the local lumberyard. Not too bad for short-order work that met my specific requirements exactly.

Of course, I could lower these time figures by keeping specialized machinery set up for sash work. If I did, my shop rates would be higher. I'd rather keep my capital expenses low and have more hourly income.

I use white pine for all my sash because it strikes a good balance between machinability and durability. Straight-grain, knot-free wood is essential because the thin, narrow muntins need to be as strong and stable as possible. Also, the outer frame can twist if it's made from wood with unruly grain. I try to use all heartwood, which is stronger and more rot resistant. It's best to cut all the rails and stiles from parts of the board that have vertical grain. These quartersawn lengths of wood are less liable to warp and twist.

John Leeke is an architectural woodworker in Sanford, Maine. Photos by the author, except where noted.

To replace old frames, first remove the exterior casing, as shown above, then remove the interior casing. This exposes the casing nails that hold the jambs to the rough framing. These can either be pulled or cut with a hacksaw blade.

Two kinds of sash—Here I'll describe how I made double-hung sash for jambs that don't have parting strips. So the sash shown in the photos don't have weather stops. *(Parting strips, weather stops* and other sash terms are explained on the facing page.) However, many older sash are made for jambs that have parting strips, and so require meeting rails with weather stops, as shown in the drawing. Meeting rails with weather stops are thicker in section and narrower in elevation than ordinary rails, and are mortised to receive tenons on the stiles, rather than the other way around. If you have to make meeting rails like this, the joinery is the same, except that the stiles are tenoned and the rail is sized to overlap (with bevel or rabbet) the other rail. You can avoid this trouble altogether, if you wish, by applying the weather stops (with brads and waterproof glue) after the sash are assembled. The instructions that follow are for simple sash.

Sequence of operations—I did all the work on this job with ordinary shop equipment—a table saw, a drill press and a router, which I mounted on the underside of my saw's extension wing. The techniques described in this article can be adapted to produce sash in new construction, casement windows and fixed-glass windows.

I begin by disassembling one of the old sash to determine how it went together, and to get familiar with its decorative and structural details. Then I measure the inner dimensions of the old jamb, and make a drawing of the sash that shows the important features. From the drawing I compile a list that itemizes the parts and tells the dimensions of all the separate pieces.

The sequence of operations in the shop goes as follows: I thickness-plane all the stock (this can be done on the table saw since all the members are fairly narrow), and then cut the tenons and copes on all of the rails and horizontal muntins. Next, I cut the mortises in the stiles and vertical muntins. After this I set up to mold the inner edges of all the frame members on their inside faces, and then I rabbet the same pieces on their outside faces (for glazing). At this point I usually frank the tenons on the rails. Finally, I assemble the sash.

Measurements, drawings, cutting lists—After I take out the old sash, I clean off paint buildup and dirt. If the stiles of the frame are not parallel, I size the sash to the widest measurement and allow a little more time for trimming during installation. If the overall dimensions from sash to sash vary less than ¼ in., I make all the new sash to the largest size and then trim down those that need it after assembly. If the variation is more than ¼ in., I plan to make more than one size of sash. Too much trimming can weaken the frame members.

You can usually make a good guess about the joinery of the original, but if you're doing a precise reconstruction you have to take one of the sash frames apart so you can measure the dimensions of its tenons and mortises.

On my first sash projects I made complete drawings to keep the various parts and joints

Photo: Andrew Edgar

Sash anatomy

A basic sash for a double-hung window consists of an outer frame and an inner framework of smaller members that hold the separate panes of glass. The outer frame is made of vertical members called *stiles* and horizontal members called *rails*. The bottom rail on the upper sash and the top rail on the lower sash are called *meeting rails,* and these are often made to interlock when the windows are closed. This interlock can be a mating pair of bevels or an overlap (see section drawings below), and it helps keep out cold drafts. *Plain-rail sash* have meeting rails that lack the interlock feature; their meeting rails simply abut one another. The lower rail of the bottom sash has to be beveled to fit flush against the sill, which should slope toward the outside to shed water.

In the best construction, frame members are held together by wedged through mortise-and-tenon joints; as a general rule rails get tenoned, stiles get mortised. In some traditional sash, though, the meeting rails are rather narrow and so are mortised to house tenons cut onto the stiles. You can use slip joints, but these lose much of their strength if the glue in them fails, whereas wedged through tenons hold firmly even without glue.

The members of the inner framework or grid that holds the glazing are called by several names. I call them *muntins,* though they're variously known as *mullions, sticks, sticking, glazing bars* or just *bars*. Like the outer-frame pieces, the muntins should be tenoned into the rails and stiles, and into one another.

All the frame members—rails, stiles and muntins—are molded on their inner faces and rabbeted to hold glazing on their outer faces. This arrangement requires that tenon shoulders be made to conform to the molded edge of the mortised member.

In traditional sashmaking, there are two ways to shape the tenon shoulder. The first method involves cutting away the molded wood and shaving the shoulder on the mortised piece flat to receive the flat shoulder of the tenon. This means the beads are mitered on both members. The second, and easier, way is to cope the tenon shoulder. Simply stated, a cope is a negative shape cut to conform precisely to the positive shape that it fits up against. —*J. L.*

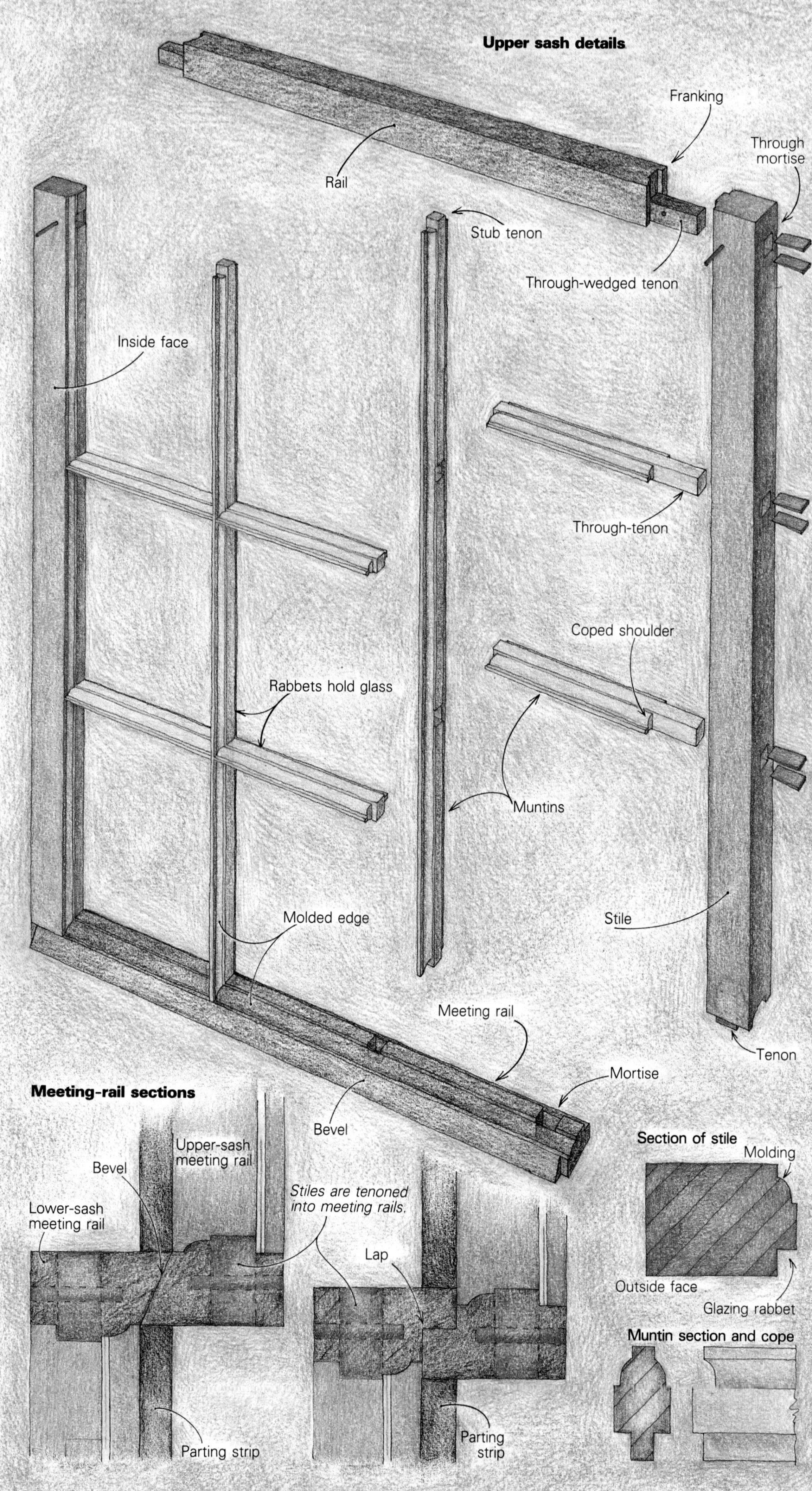

Illustrations: Christopher Clapp

Tenoning with a router. **With the router mounted on the underside of his saw's extension wing, Leeke removes the correct amount of wood to produce the cheek of a tenon. The stock is fed into the bit with a push gauge; it squares the workpiece to the cutter and keeps the wood from splintering out on the back side of the cut. A shop-vacuum hose pulls chips through a hole in the fence.**

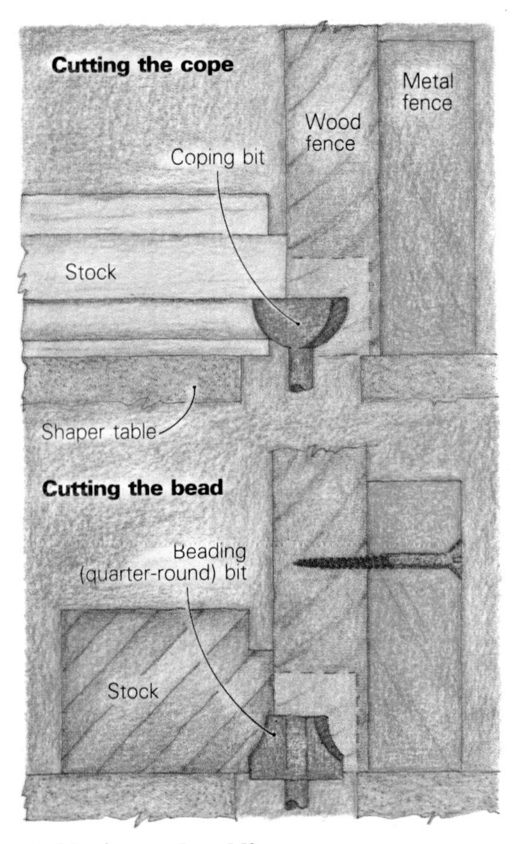

Ripping muntins to width **from stock that has already been tenoned and coped saves time and minimizes splintering and tearout from cross-grain cutting. Because the horizontal muntins are thin and short, a pair of push sticks has to be used.**

Rabbeting and molding **muntin edges requires the use of two hold-downs. The one attached to the fence holds the stock against the table, and the one clamped to the table holds the work against the fence. Kerfs in the hold-down blocks allow the wood to spring and flex against the stock.**

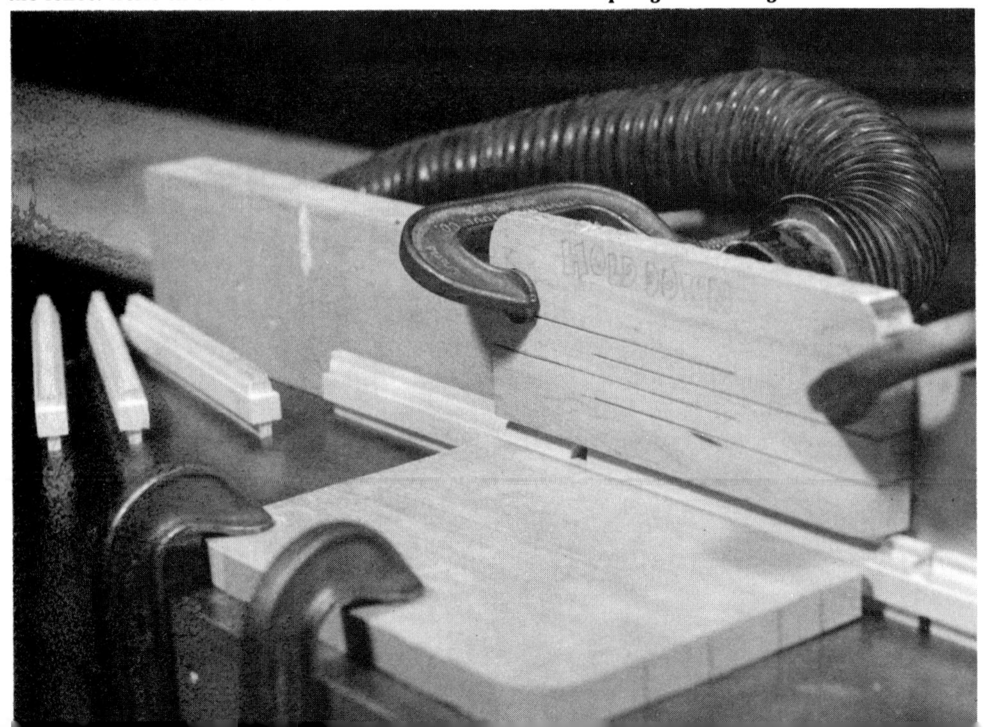

straight in my mind. From these drawings, I made a list of all the pieces I'd need. Each part has its own line, and each line is keyed with a letter to the corresponding part on the drawing. Also on the line is the size, the quantity needed, the name of the part, and its location in the sash.

Router-table joinery—In millwork shops and sash-and-door plants, tenons are cut by single-end or double-end tenoning machines, which cope the shoulder of the joint at the same time they cut the tenon. In smaller custom shops, spindle shapers do the same job. But these are expensive machines, and they take up a lot of space, which I don't have much of. And I was in a hurry. So on this job I improvised a router setup to cut the tenons and copes.

Instead of building an extra table in my already crowded shop, I mounted a router under one of my table-saw extension wings. This arrangement lets me use the saw's rip fence and miter gauge as working parts of the router setup. To keep from drilling needless holes in your saw table, be careful when you lay out and bore the hole in the cast iron for the arbor and cutter and the tapped holes that will let you attach the router base to the underside of the wing with machine screws.

I made a wooden auxiliary fence 2 in. thick and 5 in. high to attach to my saw's metal fence. To suck up chips and dust from around the cut, I bored a hole through the fence and carved a socket to accept the end of my shop vacuum hose (photo top left). This keeps the chips from clogging in the bit during cutting and from building up against the fence.

To guide the workpiece, I use a rectangular push gauge made from a block of pine. By holding the edge of the block against the fence, I stabilize the stock, and square it to the line of cut. Also I notch each corner of the push gauge to each size tenon and to the copes. This way the block backs up the workpiece and keeps the cutter from tearing and splintering the wood as it leaves the cut.

Cutting tenons on rails and muntins—For plain-rail sash there are two tenon lengths—the long through-tenons on the rails and on the stile ends of the horizontal muntins, and the short tenons on the vertical muntins and on horizontal muntins where they are joined to the vertical muntins. It could be that your sash will have a third tenon length for vertical muntins that are joined into the rails, and a fourth tenon length if your stiles are to be tenoned into meeting rails with weather stops. Before setting up to cut tenons, the stock for the rails, stiles and muntins must be surfaced to final thickness and crosscut to finished length. But the stock for the rails and muntins should not at this point be ripped to final width, especially the muntin stock. It's easier and safer to cut the tenons and do the coping on wide boards; it saves time and avoids tearout as well. Remember to mill up some spare pieces for trial fitting, and to be substituted if you ruin good ones. And it's a good idea at

From *Fine Homebuilding* magazine (December 1983) 18:72-77

this point to set up the hollow-chisel mortiser in your drill press because you'll need to cut some mortises in scrap to test-fit the tenons.

Most tenons are slightly offset from the center of the stock, but because all the framing members are the same thickness, you can set the router bit to cut the tenon cheek on the inside faces of all the pieces, then reset the bit to cut the cheeks on the outside faces. Mark out the dimensions of the two lengths of tenons on a pair of test pieces, and set the bit at the precise height for cutting the face side. This requires careful measuring, for which I use a vernier height gauge.

Calculating tenon length is complicated by the need to cope the tenon shoulder, which in effect lengthens the tenon. This added length equals the depth of the molding profile, and has to be deducted from the length of the tenon. Say your stile width is 1¾ in. and your molding-profile depth is 3⁄16 in.; your through-tenon length before coping will be 1 9⁄16 in., so you'll set the fence 1 9⁄16 in. from the farthest point of the bit's cutting arc.

Once the bit is set at the proper height to cut the cheek on the inside-face side of the pieces and the fence is set to cut the longer tenons, you can begin cutting. It's best to make each cut in several passes, even if you're using a large (½-in. or ⅝-in.) carbide-tipped straight bit. You'll get better results without putting an unreasonable demand on the router's motor. Make certain when you make the final pass that the end of the stock is pressed firmly against the fence and at the same time held snugly against your push gauge. Holding the stock this way ensures that the tenon will be the correct length and that the shoulder will be perfectly square.

After the first series of cuts on the rails and on the stile ends of the muntins, you need to set up to make the first cuts on the muntins for the short tenons. To keep from moving the fence and having to set it up again when you cut the cheeks on the outside face of the rails, you can thickness a scrap piece and clamp it to the fence to shim it out from the bit's cutting arc to make a tenon of the correct length.

Once you've made the cuts on the inside face of the muntin stock, you're ready to complete the tenoning by cutting the wood away on the outside face. Leave the shim clamped in place, and reset the router bit to the proper height above the table to make the next cuts. Careful measuring here is critical because your tenons won't fit if the bit is set at the wrong height. So make a cut on one of your spare pieces and trial-fit it in the test mortise. Once you get the bit set correctly, run all your muntin stock through. Then unclamp the spacer from the fence and make the cuts on the outside face of your rail stock and the muntin stock that gets long tenons.

Coping the shoulders—Coping with a router means you have to pattern-grind a matched pair of bits—a concave bit to cut the molding on the inside edges of sash members, and a convex bit to cut the cope on the tenon shoulders (photo above right). The positive and negative shapes of the pair must be perfectly complementary or your joint won't close properly, and will have gaps. The sidebar at right explains how to grind stock high-speed steel bits to get a matched set.

Now you're in a dilemma because you need a molded, rabbeted and mortised stile to test-fit the pieces you've tenoned and are getting ready to cope. The best choice here is to rip a stile to finished width, set up the router to mold the inner edges and rabbet the outer ones according to a full-size drawing of the stile in section. But you're having to perform an operation out of its logical sequence, and that can seem a waste of time. You'll also need a couple of muntin pieces; so rip a couple to width and mold and rabbet them at the same time you do the same to the stile. Whatever you do, don't throw your sample pieces away once you've gone to the trouble to make them. If you ever need them again, you'll save several tedious hours of trial-and-error setup if you have these samples to refer to.

Now that you've got a stile prototype and a couple of muntin samples, chuck the coping bit in the router and set the height so that the top of the cutter just lightly touches the bottom edge of the tenon (stock held inside face down on the table), as shown in the upper drawing on the facing page. Next set the fence to cope the shoulders of the muntins with short tenons. Be conservative when you set up. Make a pass into the cutter, and trial-fit the piece. If the fit is bad, adjust the fence cautiously and try again. Keep making minute adjustments until you get it right. You'll get some tearout on the exit side of the cut because the edge has been molded, but this won't matter for the test piece. Now cope all the shoulders for the short tenons of the muntin stock. Next reset the fence to cope the shoulders of the long tenons.

At this point you should rip the muntins to width (middle photo, facing page). To keep from having to clean up the sawn surface with a plane, I use a sharp planer blade in my bench saw. Because the muntins are thin, I use a pair of push sticks to feed the stock into the blade. Also at this time you should rip the rails and stiles to final width.

Mortising stiles and muntins—I use a ½-in. hollow-chisel mortiser that I keep sharp for this job. You can buy one of these attachments for your drill press at most woodworking-machinery dealers. Be sure to buy a little conical grinding stone that keeps the chisel sharp. A dull chisel will tear the wood on the walls of the mortise and cause nasty splitting out on the back side of the stock. Even a sharp chisel will do some tearing out if you don't back up the cut with a maple block. I use an aluminum plate that's cut out to the precise ½-in. square dimensions of my hollow chisel. The edges of the square hole give positive support to the wood as the chisel exits, and prevent splintering altogether.

To get precise results, each mortise should be laid out with care. It's best if you use a mortising gauge and striking knife, but a sharp

Pattern-grinding router bits

You have to be able to grind your own router-bit profiles if you want to reproduce moldings or make new ones of your own design. For one set of window sash that I was making, I needed a matched pair—a quarter-round bead and a cope. To make such a a pair, I begin by buying "blanks"—common high-speed steel bits, large enough to yield the finished shape I need. For these sash I used a rabbeting bit for the molding cutter and a rounding-over bit for the coping cutter.

To lay out the shape for the cutting edge, I first coat the back face (opposite the beveled face) of each bit with machinist's blue marking dye, and then scratch the profile that I want on the surface. Dye makes the pattern easy to see when grinding. If I am reproducing a molding, I lay a thin cross-sectional piece of the molding on the back face of the bit and scratch around it.

Holding the bit in a hand vise, I grind away the unwanted steel on my bench grinder. As I near the mark I make certain to match or slightly undercut the original bevel. All of this grinding is freehand, steadied a bit by using the grinder's tool rest. For tight inside curves, I use a small, thin grindstone mounted on the work arbor of my lathe. I am careful to remove the same amount of steel from both flutes so the balance of the cutter is maintained.

After I've ground right up to the mark, I refine the shape by hand with medium and fine India slipstones, sharpening the edge in the process. Then I touch up the edge with hard Arkansas slipstones. Finally I mount the cutter in the router to test the shape it will make on a scrap of wood. If the resulting shape is not correct, I grind a little more, resharpen and test again.

To make a coping cutter match another bit, I use the bit itself as a pattern for the layout. When the grinding on the coping cutter is approaching the final shape, I slip both cutters into a simple jig that holds the shanks parallel. The jig is just two ¼-in. holes drilled in a block of hardwood with a drill press. I turn the bits so their cutting edges are next to each other. Holding the assembly up to the light, I can easily see where more steel must be removed to make the edges coincide. *—J. L.*

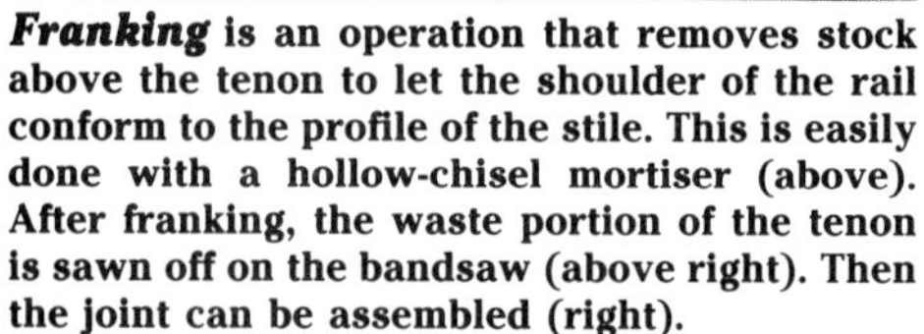

Franking **is an operation that removes stock above the tenon to let the shoulder of the rail conform to the profile of the stile. This is easily done with a hollow-chisel mortiser (above). After franking, the waste portion of the tenon is sawn off on the bandsaw (above right). Then the joint can be assembled (right).**

Assembly is straightforward. Daub the mortise walls with glue, and tap the frame members together. Then drive in the wedges for the through tenons, and snug up the frame with two bar clamps. The frame shown here is slip-joined because it will be a fixed upper unit and not subjected to the stresses that sliding sash are.

#4 pencil and a try square will do. I clamp an end stop to the drill-press table to ensure that mortises at tops and bottoms of stiles are all the same distance from the ends of the stock.

Molding and rabbeting—The next machining processes mold the inner edges of the frame members on their inside face and rabbet the inner edges on the outside face. First I install the molding bit in the router, adjust its height and set the fence. Having done this already on the test pieces, I use one of them to help with the setup. Then I mold the stiles and rails, after which I clamp hold-downs on the fence and the saw table (bottom photo, p. 64), to run the thin, narrow muntins. Molding done, I cut the rabbets, using the same straight bit that cut the tenons. Again, I clamp hold-downs to the fence and table.

Franking—When you lay out the through mortise-and-tenon joints for the rails and stiles, you'll need to leave at least an inch or more of wood between the end of the mortise and the ends of the stile, or the wood can split out. Therefore, the tenon needs to be an inch or so narrower than the rail is wide. When you cut the tenons with the router, you get a tenon the full width of the rail, so you must cut part of it away to get a tenon the right width and to get the newly exposed shoulder to fit the sectional profile of the stile. Molding and rabbeting the stile produces a proud land (flat ridge) that runs the length of the member, and the new tenon shoulder has to be relieved to accommodate this long, flat ridge.

Relieving the upper shoulder of the tenon in this way is called franking, which I do with the hollow-chisel mortiser (photo top left). What's required is mortising back behind the tenon's shoulder to the width that the land is proud and to a depth that stops at the line of the top of the finished tenon. Then all you do is saw away the waste (photo top right).

Assembly—I really enjoy this part of sashmaking, when all the work finally pays off. Traditionally, joints were put together with thick paint between the parts. I suspect the sashmakers in those days expected the paint to seal moisture out of the joint rather than act as glue to hold it together, but they didn't rely much on the adhesive strength of either paint or glue. They pegged the tenon through the inside face of the stile and wedged it top and bottom from the open end of the mortise.

Most weatherproof glues seem to work well, except for formaldehyde-resorcinol, which can bleed through to the surface after the sash is painted. When all the parts of a sash are fitted together, I snug up the joints with a bar clamp, and wedge the through tenons.

After wedging I check the sash for squareness and then peg the joints. Pegging is especially important for the joints of the meeting rail on the lower sash, because they are subject to a lot of stress during use. If pegs pass through to the outside of the sash, they could let water get into the joint. My pegs stop just short of the outside surface. □

Installing sash

When you're deciding what to do with sash that are in poor shape, you should consider the window as a whole. Examine the frames for deterioration. Windows on the south side of a house are subject to repeated wetting and drying. This can cause large checks or cracks in sills and stiles, so that these members need to be replaced. The north side of the house will be damper because it is in the shade, and you're likely to find rot in the joints of the sill and the jamb stiles.

Jamb and casing—If the sill or jambs need replacing, I usually take the whole window frame out of the wall. I begin by carefully removing the exterior casing and moldings, using a prybar (see p. 46) to lever them from the wall. This exposes the space between the jamb stiles and the structural framing (trimmer studs, header and rough sill), which gives me room to cut the nails that hold the frame in place. I use a reciprocating saw or hacksaw blade for this. There also may be nails that hold the interior casing to the frame that will have to be cut.

Once the frame is loose, it can be pulled out at the top and removed from the wall. I try to use heartwood for replacement parts of the jamb and casing because it is more rot-resistant than sapwood. In any case, the parts should be treated with a water-repellent preservative before assembly and reinstallation. Reusing old casing and molding that are still in good shape helps blend the new work in with the rest of the house. If the old jambs are still good, I scrape and sand their inside surfaces so they are flat and smooth for the sash to slide against.

Fitting the sash—To install double-hung sash with the upper sash fixed in place and the lower sash sliding up and open, you first trim the stiles to fit, then glaze and finally install the sash in the jambs.

Start sizing the top sash by planing the edges of the stiles so the sash will fit into the frame without binding. This should be a free-fit, but not so loose it will rattle side-to-side. Put the sash in place and slide it up to the frame header. If the top rail of the sash doesn't fit uniformly flush along the header, it should be scribed to the header with dividers and trimmed to fit. Then fit in the two outer stops, which are strips of wood that lay flat against the jamb stiles and hold the upper sash in place. If the top rail of the bottom sash and the bottom rail of the top sash are made to overlap and form a weather stop, your jambs must be fitted with parting strips. These are strips of wood that are let into grooves, one in each inner face of the jamb, that run the length of the jamb stile and serve to separate the two sash so they don't slide against one another. Fit the parting strips into the grooves in the stiles so they are held in place by compression only. Don't glue or nail them in place.

Next trim the lower sash to fit by planing its side edges until it runs smoothly up and down in the frame. Set the sash in place with the bottom rail resting on the sill. Then scribe the bottom rail to the sill with dividers set to the distance between the top surfaces of the meeting rails. Plane off the bottom rail to the scribed line (photo above right), forming a bevel that matches the slope of the sill. The weather stops should fit tightly together when the bottom rail of the lower sash is against the sill. If too much is planed off the bottom of the lower sash, this fit is lost. So take some care when trimming for this fit.

When the sash are sized to fit, they should be treated with a water-repellent preservative and primed for painting. Do not prime the side edges of the sash. They should be left bare to slide against the stiles.

Glazing, painting and finishing—I usually take sash to the glass shop to be glazed. The glass should be bedded in a thin layer of glazing compound and set in place with glazing points. When complete, the glazing compound should have a neat beveled appearance and not show from the inside.

The sash should have two top coats of paint. I prefer oil-base paints. Run the paint just slightly onto the glass, thereby sealing the glazing from rainwater. Do not paint the edges of the sash that will slide against the frame. When the paint is dry, wash the glass.

To install the window in its jamb, set the top sash in place and slip the parting strips into their slots. Trim any beads of paint that may have dried on the side edges of the lower sash, and test to see if it still slides freely in the frame. When you are satisfied with the way the sash fits, then secure it in place with the beaded or molded inner stop, taking care to use thin brads so as not to split the wood. These stops should be carefully positioned so the sash is free to move but not so loose that it will rattle in the wind. —*J. L.*

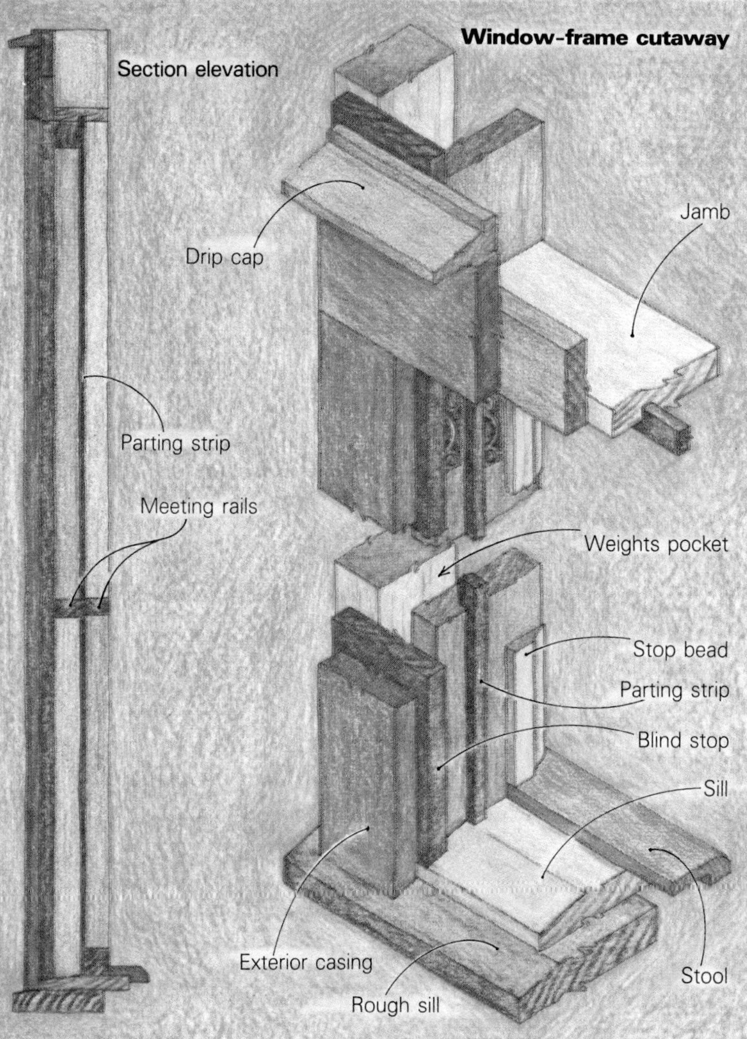

Photo this page: Andrew Edgar

Making Curvilinear Sash

How to lay out and assemble a semi-elliptical window

by Norman Vandal

In fine Federal and Greek Revival structures, curvilinear windows are common—above pilastered entrances, in gable pediments and incorporated with Palladian windows. Built under roof gables or in other confined areas where conventional rectangular windows couldn't fit, the curvilinear window was a decorative means of providing light to upstairs rooms and attic space. Since I'd recently been commissioned to build such a window for a nearby restoration project, I began to notice many more of the type I'd been asked to create. I became critical, noting some of the small details that separate the fine from the not-so-fine. Fortunately, there are many semi-elliptical windows to be found around Roxbury, Vermont, where I live. I was able to examine different styles and to develop a taste for the most desirable features.

There are probably many ways of producing curved sash, but I could find little information on building this type of window by hand. Avid old-tool enthusiasts, concerned with the function of the tools in their collections, seem to be doing most of the research. As a rule, most craftsmen rely on large millwork companies to produce on the assembly line the items they need, although cabinetmakers sometimes stumble into sash work when making such pieces as corner cupboards or secretaries.

Drawing the semi-ellipse—This is the first step in making curvilinear sash, and will determine the proportions of the window. I learned how to draw an ellipse from Asher Benjamin's

Curvilinear windows were popular features in houses built during the Federal and Greek Revival period. Even today curved sash must be made largely by hand, using traditional techniques of fine joinery.

The American Builder's Companion, an 1806 guide to neo-classical detail and proportions, reprinted by Dover Publications (180 Varick St., New York, N.Y. 10014) in 1969. The method I find most valuable for making a full-scale template is illustrated in figure 1. Using this technique, one can draw concentric ellipses, a necessity for making templates. Note, however, that the resulting curve merely approximates an ellipse, since the points *d* and *d'* on the major axis are not foci in the usual sense.

The distance between *a* and *b* will be the maximum length of the bottom rail (in my window, 3 ft., with a height of 13½ in.), but you must draw the entire ellipse to be able to locate the compass points from which the top of the curve is described. The window I made (photo left) is rather elongated in comparison to many old windows I have seen. You can change the proportion of your ellipse simply by experimenting with the compass points *d* and *d'*. Moving them closer to the center point in equal increments will elevate the semi-ellipse by the amount of one increment. With a little experimentation, you'll eventually reach a pleasant proportion.

Making the templates—After determining the proportions of your semi-ellipse, begin to construct the templates. For these I use poster board, available in art-supply stores. Try to get board as large as the full-scale sash to avoid taped seams on the template. You'll also need a piece for exterior and interior casing templates.

To construct the templates, you'll need a compass capable of scribing large-diameter circles. For smaller sash, a pair of large dividers will suffice, though large dividers are difficult to locate. You can make dividers of wood, using a nail and a pencil as points and a wing-nutted bolt to secure the joint—a little crude, but functional. My beam compass, an old set of wooden trammel points fastened to a wooden beam, is shown in figure 2. One point is permanently attached; the other, which slides along the beam, can be set in place with a wooden wedge.

Begin by compassing your exterior sash dimensions on the poster board. Then determine the width of the rail (mine generally measure

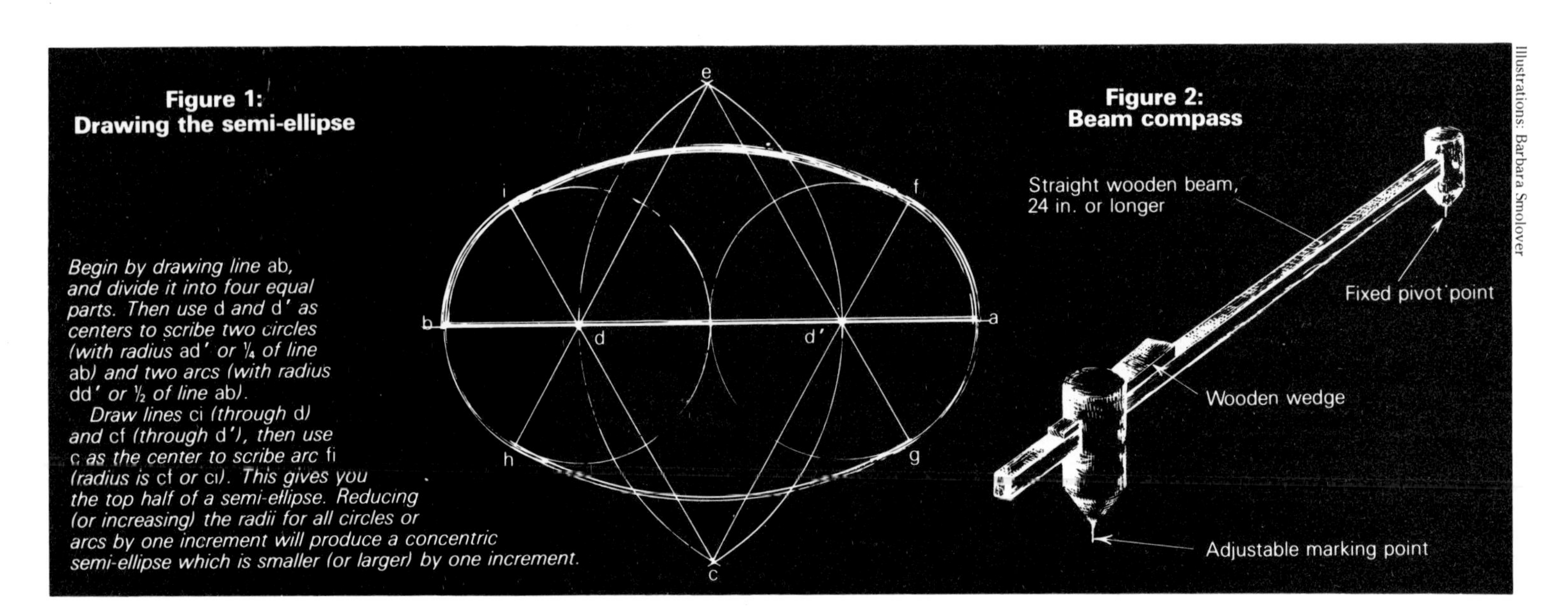

Begin by drawing line ab, and divide it into four equal parts. Then use d and d' as centers to scribe two circles (with radius ad' or ¼ of line ab) and two arcs (with radius dd' or ½ of line ab). Draw lines ci (through d) and cf (through d'), then use c as the center to scribe arc fi (radius is cf or ci). This gives you the top half of a semi-ellipse. Reducing (or increasing) the radii for all circles or arcs by one increment will produce a concentric semi-ellipse which is smaller (or larger) by one increment.

Illustrations: Barbara Smolover

From *Fine Homebuilding* magazine (October 1981) 5:25-27

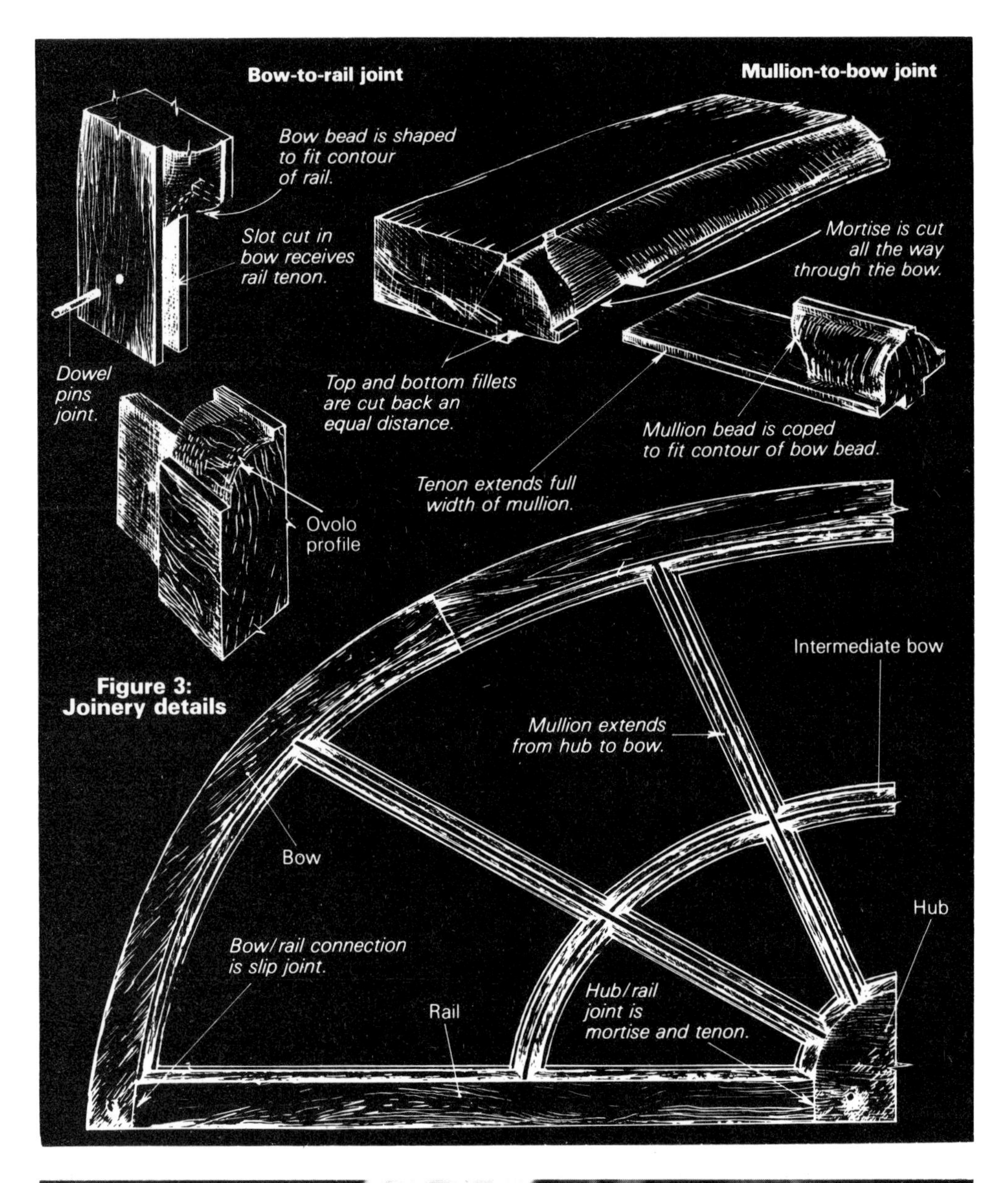

Making the mullion-to-bow joint: Once the intersection angle has been marked out on both mullion and bow, the mortises can be drilled out and squared with a chisel. Then the tenon is cut in the mullion and test-fitted in the mortise so the curve can be scribed in the bead, as shown. Use a coping saw to cut the curve and a rattail file or an in-cannel gouge to shape the joint for a snug fit.

1½ in. to 1¾ in.) and the width of the bow. Draw the line representing the top of the bottom rail parallel to the line *ab*. To draw the inner contour of the bow, you must reduce the radii of the circles with the centers *d* and *d'* by the exact width of the bow.

To avoid the disadvantages of short grain, it is best to make the bow out of three pieces of wood instead of one, orienting each to maximize long grain. Draw the lines *ci* and *cf* on the template to indicate the position and angle of the joints on the bow. Later, when you're cutting up the template, you can divide the bow into three segments. The sash shown in figure 3 has an intermediate bow, something you may want to avoid in early attempts at making curvilinear sash. If you decide to use an intermediate bow a short distance out from the hub, make it concentric with the curve of the primary bow. An intermediate bow will have one more segment than the number of mullions.

You need not cut templates for the mullions, as they are of straight stock and will be cut to proper length later. But you'll need to decide how many mullions you want in your sash, and to determine what angles they form with bow and hub sections. At this stage you may want to consult old work—the joinery details you are able to uncover in a traditionally made window will prove helpful from this stage onward. In terms of workability and historical accuracy, mullion width should be between ⅝ in. and ¾ in. This will equal twice the width of the bead plus the width of the fillets.

The hub, where the mullions will converge, can be either semicircular or semi-elliptical, but make it large enough to keep the rays from coming into contact with each other, thus keeping the joinery less complicated. The hub must be cut separately, mortised to the rail, and then mortised to receive the mullions.

I usually make my templates ⅛ in. longer than their finished dimensions. You need this extra length when you cut and fit the joints, and any excess can be trimmed off during final assembly. When you have completed the sash templates, make templates for the exterior and interior casings. My exterior casing, which overlaps the sash ½ in. to keep it from falling out, is in two sections, divided by a keystone. Cut out the templates with a stencil knife. I use a surgical scalpel with sharp blades to keep the edges crisp.

Cutting the parts—Your cardboard templates will serve as patterns for laying out all the parts to be cut. I use clear 5/4 eastern pine stock, although mahogany, clear spruce or basswood would serve equally well. Curved cuts for bow sections and the hub can be accomplished with frame saw, bandsaw or saber saw, depending on your equipment and inclination. Remember to cut the mullions a bit long, since some trimming is inevitable with the type of joinery you'll be doing. To join the bow segments, I use a simple butt joint reinforced with dowels—two to each joint—and yellow wood glue.

During the Federal and Greek Revival periods, a complete set of highly specialized tools available from tool-makers and woodworking suppliers provided the joiner with everything needed to make all types of sash. While many of these tools are still around today, most are useful in dealing only with square or rectangular sash. One elusive tool is the sash shave, a cross between a spokeshave and a plane; without it curved sash is practically impossible to make by hand. I could not locate an old sash shave to match any of my planes, so I began to make one from scratch. I quickly realized it was going to take more time and research than I had anticipated (I still haven't finished it), so instead I used a router to cut the molding and the rabbets. You can also use a shaper. The only bits you'll need are a simple ¼-in. bead (with integral stop) and a ¼-in. rabbet (also with integral stop). The bead cutter gives an ovolo molding profile (figure 3), the easiest shape to work with when coped mortise-and-tenon joints have to be cut.

Mortise-and-tenon work—This is where the craftsmanship comes in. The two bow-to-rail joints are the first ones to make. Here you have the option of either mortising a bow tenon into the rail, or cutting a slip joint. I chose the latter because part of the slot and tenon-cutting work could be done on my radial arm saw. The mortise is more secure in terms of joint movement. The procedure for making all sash joints is the same: Lay out the joint first by aligning the parts, mark the stock, and then cut the tenons and mortises. I cut my mortises all the way through, since this makes squaring the cavity easier (you can chisel in from both sides). Check and adjust these for fit, mark where shoulder areas must be coped, and then cope the contoured part of the joint. This is by far the toughest part of the joinery work, and must also be a "check and adjust" operation. I use an in-cannel gouge (a curved gouge with its bevel cut on the inside) and a rat-tail file to shape the curved shoulders of the joint. You may want to use a coping saw to cut the cope into the mullions, but only a gouge can be used on the hub and bow-to-rail joints. Final shaping will consume the extra ⅛ in. or so added to finished length when the parts were first cut.

Once the bow/rail joints are secure, join the hub to the rail (temporarily) and lay out the mullions on the bow/rail/hub assembly to mark where mortises have to be cut. Fashioning the mortises and tenons for mullion-to-rail and mullion-to-hub joints can be made easier if you remember that the tenon extends the full width of the mullion and is made by cutting away the bead or contoured part of the mullion and the bottom fillet. To make the mortise, drill and chisel out the square central portion of the bow or curb between the bead and the rabbet.

Once the mortises and tenons have been cut and fit smoothly, you have to cope the curve in the mullion bead which completes the joint. First seat the mullion tenon in its bow (or curb) mortise, then mark the shoulder cut on the mullion bead, as shown in the photo on the previous page. This has to be done by eye. You can expect some trim work with gouge or file (after you make the cut with a coping saw) to achieve a flush joint, but it's surprising how tight a cope you can cut with a little practice.

When all the mullions have been cut and fitted, a trial assembly is in order. Slip the tenons of the mullions into the bow, and snug up the rail to the bow as you're fitting mullion tenons into the hub. With luck, everything will fit as you planned, but don't be alarmed if a little further trimming is necessary. When you are satisfied with the results, glue up all the joints and reassemble the sash. A peg glued in the two bow-to-rail joints will provide extra strength.

Making an elliptical sash without an intermediate bow, as described above, is the easier and more advisable route if you've never built a curved sash before. If you're incorporating an intermediate bow, as I did, then these bow sections would be cut and mortised into the mullions before final glue-up of the sash.

The jamb—Since all of the old curvilinear window units I observed were in place, I was not able to examine the way in which the jamb units were constructed and framed into the walls. I had to work out my own system, shown in figure 4; you may wish to do likewise. Total wall thickness, sheathing material and stud spacing will determine the width of the jamb and the framing details.

Before you can make the jamb unit, first make the window stool to support the casings. I use 8/4 native white pine, and bevel the stool 10° toward the outside to shed water. (You also have to bevel the bottom of the rail so the rail-to-stool joint is sound.) The stool should be longer than the rail if you want ears at each end. If ears are not part of the design, trim the stool flush once you've attached the jamb.

Although you can make the stool any width, extending it too far increases the amount of weather it will have to endure. Drip kerfs (grooves on the underside of the exterior edge of the stool) allow water to drip free from the building's face. A groove cut to accept the clapboarding under the stool is an aid to both fitting the siding and weatherproofing the unit.

The sides of the jamb unit can be made from two pieces of exterior-grade plywood—I used ¾-in. thick pine stock, although ½-in. plywood will be strong enough if you use more blocking. The elliptical opening in each panel should be a full $\frac{3}{16}$ in. greater than the final dimensions of the sash on all sides. Once these are cut, attach the blocking that connects both panels. Now you've got a single unit to work on. Invert it to attach the $\frac{3}{16}$-in. thick jamb sections. They have to be bent to conform to the curve in the plywood frame. I used clear pine and soaked the wood to make it more flexible before fastening it to the ply edges with glue and small brads.

The stool can be attached next, but test-fit the completed sash first—fine adjustments at sash, jamb and stool contact points are more difficult to make once the stool is in place. I used glue and dowel pegs to join the stool to the plywood, but screws will work equally well. Once this is done the exterior casing can be fastened to the exterior side of the frame. It should overlap the bow section of the sash by about ⅛ in. Several small stops, tacked to the jamb on the interior side of the bow, force it securely against the casing.

Allow ½ in. to ¾ in. on both the width and height of the wall opening to shim the jamb unit plumb and level. Fasten the unit by nailing through the exterior casing into the sheathing and frame of the building. Now the siding must be cut to fit the curvature of the casing, but don't despair. You still have the templates, from which you can fit the siding and the interior trim. □

Norm Vandal builds and restores traditional houses. He also makes period furniture.

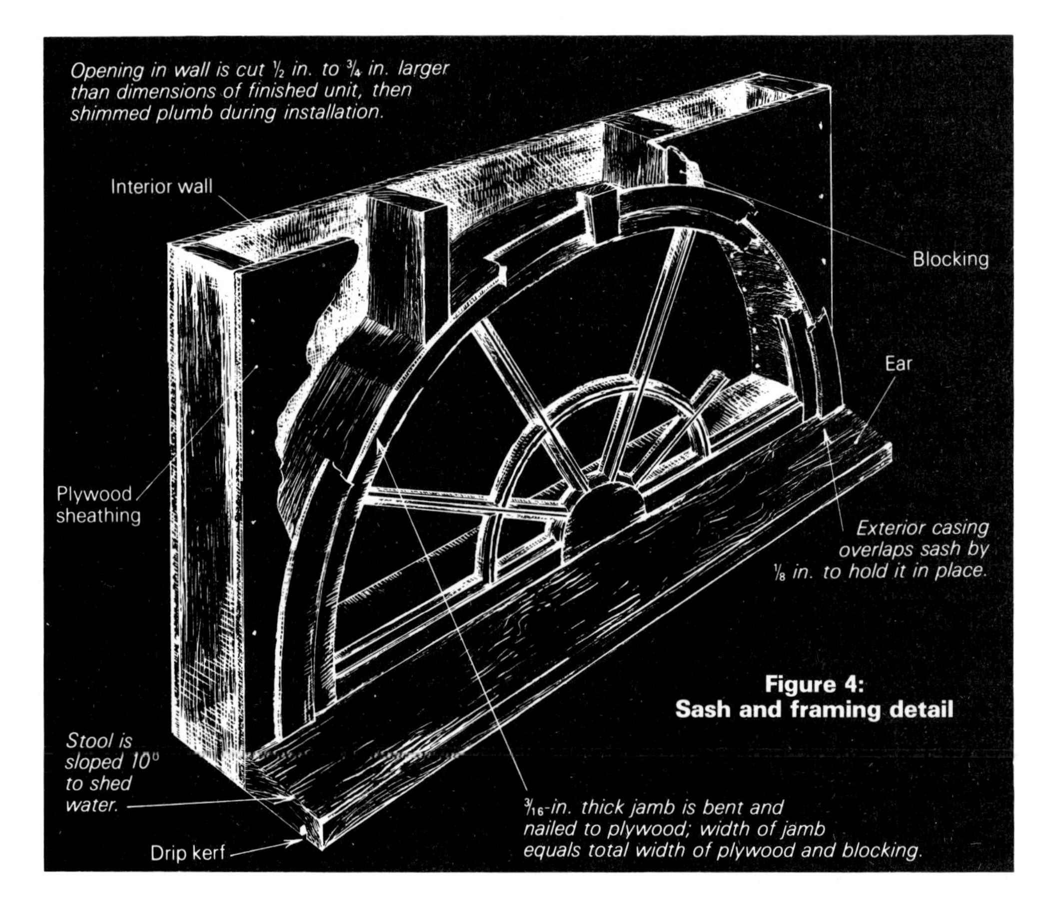

Figure 4: Sash and framing detail

Building Louvered Shutters

Jigs and careful planning make quick, accurate work of a potentially tedious job

by Rob Hunt

A few years back, I was asked to build movable louvered shutters for a Victorian house that was built in 1888. The house's new owners were restoring the building, and they wanted new shutters made to match the old ones that had deteriorated over the years. Thanks to this first commission, we've been building shutters for a number of clients. Out of necessity we've found ways to make the work go quickly and accurately.

Many early homes had shutters. Hinged to open and close over windows, and with movable louvers, they served to protect the windows from bad weather and to diffuse the incoming light. Changing the position of the louvers changes the flow of light and air through the room, giving you a range of lighting conditions to choose from. Shutters can do a lot to enhance windows, and they give you more control in adjusting the amount of daylight you want in a room, and more control over the ventilation day and night.

The frame of a shutter consists of two vertical members, called stiles, and at least two horizontal members, called rails. The shutters shown here have two central rails in addition to the top and bottom rails, dividing each shutter into three sections. The louvers are beveled slats with round tenons that fit into holes in the stiles. Each set of louvers pivots as a unit thanks to a vertical rod coupled to the louvers by an interlocking pair of staples.

Cutting the louvers to uniform size, calculating the amount of overlap (the spacing between each louver) and mounting the louvers accurately within the frame are critical steps in the production process, and we've been able to increase both the accuracy and speed of the job by using layout sticks and several jigs that I'll describe as we go along.

Layout—The first step in laying out louvered shutters is to make precise measurements of the opening where the shutters will go. Measure from the top of the window or door jamb to the sill on each side, then measure across the jamb at top and bottom and at several places in the middle. Taking extra measurements for height and width is especially important in older houses, since their jambs are seldom square.

If the jamb is out of square by ⅛ in. or less, I build the shutters to fit the smaller measurement. If the skew is worse than this, I build them ⅛ in. smaller than the largest measurement and then trim them to fit the opening after they are completed.

When you measure the height of the opening, remember that the sill is beveled. Run your tape to the point where the outside bottom edge of the shutter will hit the sill. Then

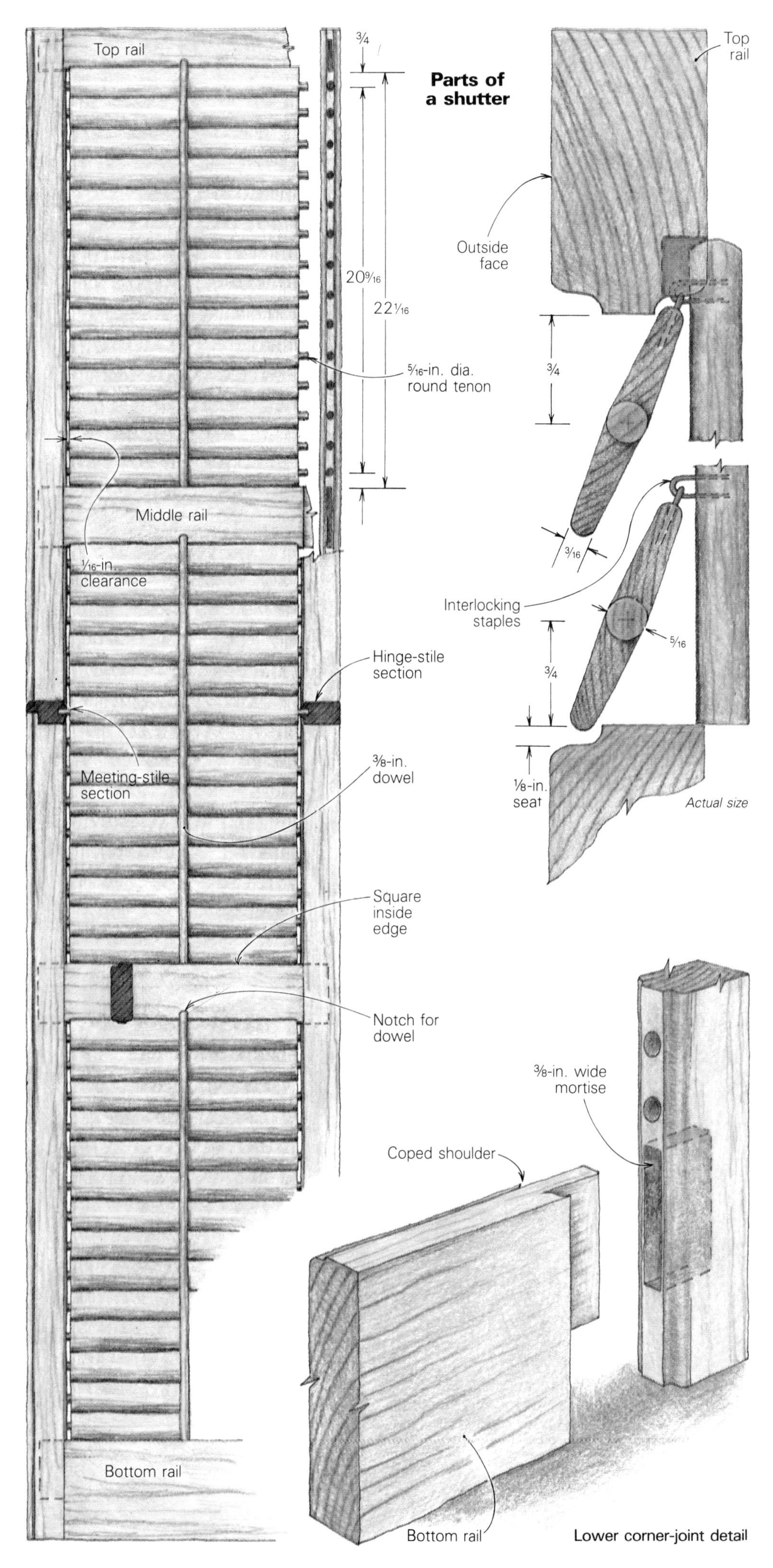

when the shutters are built, you can bevel the bottom edge of the rails.

For projects like this, which require a number of identical, precisely cut pieces, we find layout sticks very helpful. They serve as full-size templates and contain all the required dimensions and joinery details. The horizontal stick we made for these shutters is basically a full-size sectional drawing. It shows the width of the opening, the width of the stiles, the length of the rails, the louver dimensions, the bead on the stiles and the coped mortise-and-tenon joint that joins stiles and rails. All these measurements are critical, since each pair of shutters has to swing closed along a ¼-in. rabbet. And finally, the louvers need to operate smoothly, without binding.

The vertical layout stick for these shutters is marked off to show the length of the stiles, the location of the rails, and the stile holes that accept the tenoned ends of the louvers.

The spacing of the louvers is determined by louver size and the degree of overlap desired, the overall height of the shutter and the width and number of rails per shutter. The shutters shown here are 79¾ in. high and have four rails. The top rail is 2 in. wide; the two center rails are 3 in. wide; the bottom rail is 5½ in. wide. Measuring between the top and bottom rails gives us 72¼ in.; so to get three equal shutter bays and allow for the 3-in. wide central rails, each bay must be 22 1/16+ in. high (the plus means a heavy 1/16 in.).

The next thing to figure out is the number of louvers you need, and the spacing between them (actually the spacing between bore centers for the holes in the stiles that receive the round tenons), so that the holes for the tenoned ends of the louvers can be marked on the stiles. Louvers on most Victorian shutters are 1¾ in. wide, and this is the width we used (Greek Revival style shutters have 2¼-in. wide louvers).

The top and bottom louvers in each section must be located first, since their positions determine how much the louvers can close. Closing the louvers onto the bead is best for shedding water, but if you want to shut out the light, you need more pivoting clearance at top and bottom, which will allow the louvers to seat against each other. We centered the holes for the top and bottom louvers ¾ in. from the top and bottom rails, sacrificing complete closure for a ⅛-in. seat against the bead on the rail (top detail drawing).

Once the holes for the top and bottom louvers have been located, we can figure the spacing for the rest of them. We want the louvers to overlap about ¼ in. when closed to block the light and shed water, so the tenon holes in the stiles should be approximately 1½ in. apart, give or take a very small amount. The next thing that we do is to measure the distance between the top and bottom hole in each section (20 9/16+ in.) This we divide by 1½ in., yielding (thanks to my calculator) 13.713 spaces per section. You can't have fractions of a louver, so I divide 20 9/16 in. by 14, and get 1.469, or 1 15/32 in. between louvers. This is the spacing that I check by locating

Illustrations: Christopher Clapp

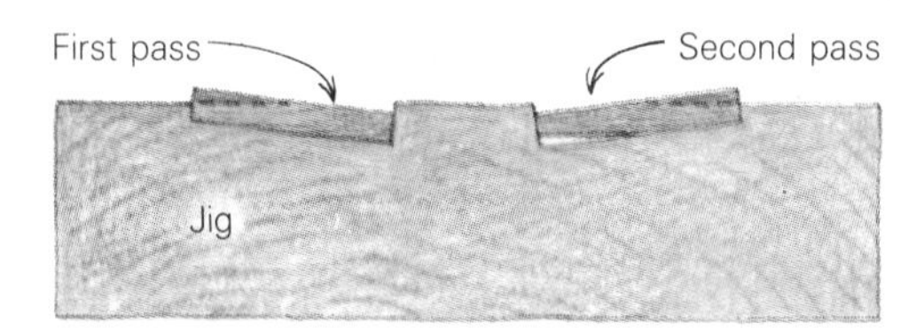

Shaping louvers. **Slats of white pine or cypress are beveled by running them through a jig (drawing above and photo at right) that's clamped to the planer bed. Grooves in the jig are angled at 6° so that the planer removes a narrow wedge of wood to make the bevel. Flipping the slat and running it through the other groove completes the beveling job. A second jig, far right, acts as both fence and stop as the squared tenons of each louver are rounded with a plug cutter. The shoulders of the tenons have already been cut on the table saw.**

the centers for the dowel holes on my vertical layout stick.

Another way to space your louvers is to adjust your rails so that 1½ in. will divide into the bay height equally. But for most of the shutters we've built, I've been matching existing shutters, so rail sizes and locations have been fixed.

Cutting louvers—With layout work done, it's time to make the parts. We start with the louvers. The completed louver needs to have two bevels on each side, and the long edges have to be rounded over. Each louver also has to be tenoned at each end to fit into the holes in the stiles.

To produce the slats for the louvers, we first take rough 8/4 stock—either white pine or cypress—and mill it down to a net thickness of 1¾ in. Then we rip strips ⅜ in. thick by 1¾ in. wide and surface 1/32 in. off of each side.

To bevel the louvers, we use a special jig, sometimes called a slave board, that we clamp to our planer bed. It's made from a piece of 1¾-in. thick oak that's slightly longer than the planer bed and about 7 in. wide. Two shallow, 1¾-in. wide grooves in the top face of the slave board are what make the jig work. The groove bottoms are angled 6° off the horizontal, and this slightly angled running surface for the wood strips lets the planer waste a narrow triangular section from the strip to create the bevel (photo top left). Each strip of wood is fed through one groove, then flipped and fed through the other, giving you all four bevels with just two passes through the planer, as shown in the drawing above.

To round the edges, we run the beveled strips through a 3/16-in. bead cutter on the shaper, using feather boards to hold them in place. Now all that remains is to cut the slats to length and cut round tenons on their ends. The shoulders of the louvers should clear the edges of the stiles by at least 1/32 in. or else they will bind when they are painted. We use a clearance of 1/16 in. to stay on the safe side.

We cut the louvers to a length that includes the ⅜-in. long tenon on both ends. Then we cut to the shoulder lines of the tenons on the table saw, holding the louver on edge against the miter gauge and setting the blade height to leave a ⅜-in. square tenon on both louver ends. The tenons get rounded with a 5/16-in. plug cutter chucked in a horizontal drill press (Shopsmith). As shown in the photo top right, we use a fence to guide the louver into the plug cutter and a stop to keep from cutting into the shoulder.

With a hollow-chisel mortiser, the stiles are mortised to receive rail tenons. Holes have already been drilled to receive the louver tenons.

Stiles and rails—We start with rough 6/4 stock, joint one edge and one face, and then surface it to a net thickness of 1¼ in. (frames for some shutters may be as thin as 1⅛ in.). The stiles for these shutters are 2 in. wide, and they're mortised to house tenons on the rails. We cut the mortises with a hollow-chisel mortiser (large photo above) and the tenons on a small, single-end tenoner.

The holes in the stiles for the louvers get drilled at this time too. Rather than mark and drill the stiles directly, we've found it more foolproof to make a drilling jig. It's the same length and width as the stile, and the hole centers are transferred onto it from the vertical layout stick and then drilled out on a drill press. We attach the template to the stile with C-clamps, then start drilling, making sure the depth of the stile holes is ⅛ in. deeper than the length of the tenons.

The lower edge of each rail on the inside face of the shutter is left square, while the upper edge needs to have a short notch in its center to accept the vertical dowel. The notch and the square edge allow the shutter to open and close securely.

The remaining inner edges of the frame get an ovolo bead. This means that the shoulders on the rails have to be coped where they meet

Assembling the shutters. **Above, test-fitting the mortise-and-tenon joints between stiles and rails before final assembly. The grooved jigs on the table will hold the louvers as the shutter is assembled, as shown at left. Rails are joined tightly to the right stile, but loosely to the left one, providing clearance so the tenoned louvers can be inserted in the stile holes. The final step, below, is stapling the vertical rod to each louver section. Double-stapling keeps louvers aligned with each other, so they can be opened and closed as a unit.**

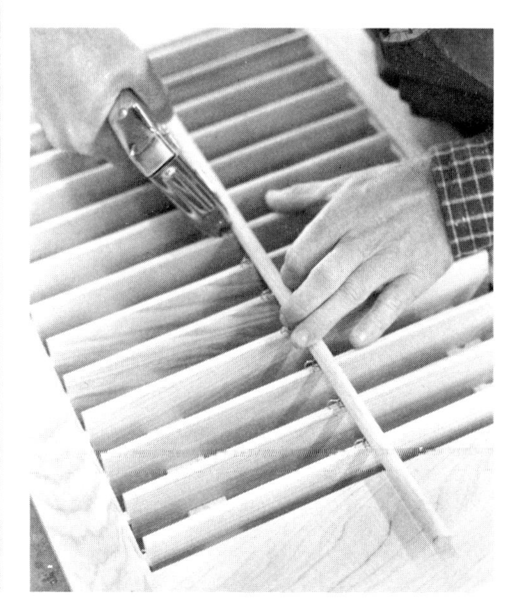

the molded edges of the stiles. Coping is tough to do by hand, so we use a shaper fitted out with a three-wing coping cutter (a Rockwell 09-128 male sash) on a stub spindle. The stub spindle allows the rail's tenon to pass over it, so we can cope one shoulder at a time.

The last pieces to make are the dowels that will be attached vertically to each bank of louvers. We make these on a shaper with a ½-in. beading bit.

Assembly—Before putting everything together, we need to make the jigs that will hold the louvers in uniform position. The drilling template for the stiles can be converted to a louver jig by cutting ¼-in. grooves across the center of the holes. We make another jig just like this one and use the pair to hold the louvers. In the photo above left, both jigs are lying on the table while I test-fit stiles to rails.

Assembly isn't really that tricky when you use these jigs, though you have to work faster than the glue that's used on the mortise-and-tenon joints (we use West Systems Epoxy, made by Gougeon Co., 706 Martin St., Bay City, Mich. 48706). We first glue and assemble one side of the shutter completely, pressing all stile-to-rail joints home. Then we lay this sub-assembly on top of the jigs, apply glue to the exposed rail tenons, and engage them in their mortises. The trick here is to leave just enough clearance for the dowels at the ends of the louvers, as shown in the photo below left. Here's where fast work is important. Have your louvers ready, get them all engaged in their holes and set in the jig, then close the joints between stile and rails.

You can cut the dowel to length after stapling it to the louvers, or you can cut before stapling. The length of the dowel is the distance between the rails plus the length of the groove in the rail at the top of the section. Round the top of each dowel so that it will fit into the groove when the louvers close.

The dowel receives staples at intervals equal to the distance between louvers; and the uppermost staple on each dowel should be located far enough down the dowel so that the dowel fits into its groove in the upper rail when the louvers are closed. We use an Arrow T25 stapler with $\frac{9}{16}$-in. staples. It's a stapler that is used for putting up small wire, so the staples don't sink all the way in.

Next we attach each dowel to its section of louvers by shooting staples into the louvers through the staples already in the rod (photo below right). We learned, after some mishaps, that grinding the bevel off the staple points made them shoot straight in, with a minimum of splitting.

We usually leave painting the shutters to someone else, but it's important to seal the wood with a wood preservative before the finish coats are applied. Spray application is far better than brush-on because of the shutters' many movable parts. □

Rob Hunt is a partner in Water St. Millworks and a cabinetmaking instructor at Austin (Tex.) Community College.

Photos: Jane Hunt

Shutters from central Texas

Bright sunlight, long days and humid, hot weather made operable louvered shutters necessities for 19th and early 20th-century Texas houses in the San Antonio and Austin areas. The louvers let air in and kept rain out even on the foulest days, and on clear, warm days, they screened interior spaces from direct sunlight, while allowing for cross ventilation and natural convective cooling. Both air flow and ambient light levels can be regulated by adjusting the louvers. As the examples here show, shutters can enhance various architectural styles.

The photo directly below is a view from inside the window bay that's seen from the outside in the photo at bottom left.

Screen-Porch Windows

Some site-built alternatives to manufactured window units

by Douglass Ferrell

These unobtrusive accordion windows fold together against the wall, but can be quickly drawn across the screened opening if a storm blows up.

Some of my best childhood memories took place on the old screen porch at my family's lake cottage. Out there around a big wooden table, we used to play cards and board games late into the night. Moths, mosquitoes and other bugs droned beyond the screens and across the dark lake. On rainy days we assembled jigsaw puzzles out on the porch and made napkin rings and other trinkets out of birch bark.

Ever since then, I have been interested in screen porches and what makes a good one. For me the key elements include wood framing, large expanses of screen, low window sills and a long wall facing the view, which might be a body of water, treetops seen from a rise or simply an expanse of green lawns.

The screen porches favored in the Sun Belt, with concrete-slab floors, aluminum framing and floor-to-ceiling screens, have the advantage of not requiring operable windows over the screens to keep out the weather. But these porches aren't suited to northern climates. The aluminum and concrete are too cold, and full-length screens forfeit an important sense of shelter.

Wooden screen porches, on the other hand, need some type of operable windows to protect both the structure and its furnishings from driving rain and drifting snow. Although the design and construction of the rest of the porch may be straightforward, the windows can be a challenge.

On any porch, a series of factory-made windows or glass doors with all their jamb and sash components cluttering up the view just can't

From *Fine Homebuilding* magazine (August 1987) 41:41-43

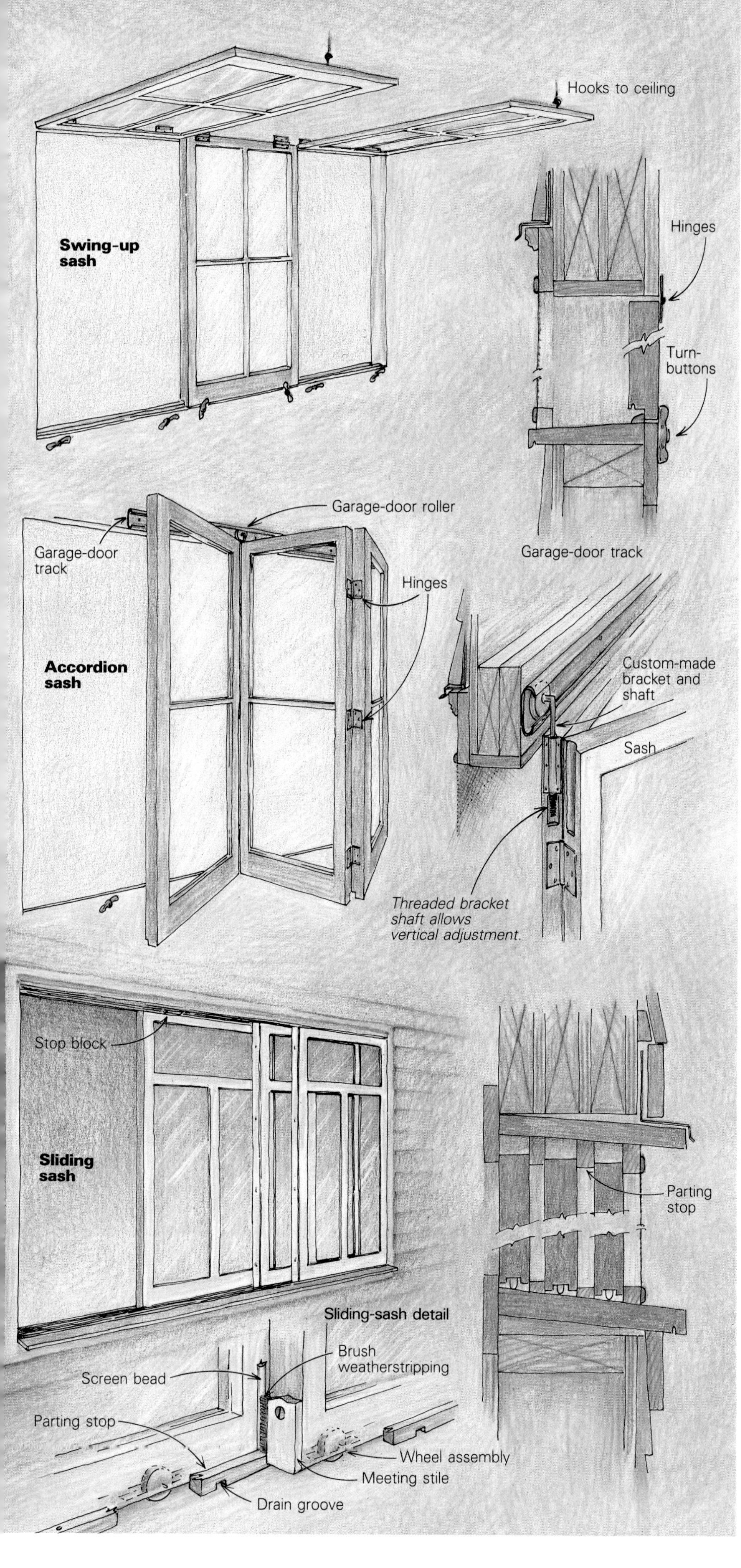

provide the airy feel of a good screen porch. In the ideal window system, the sash should store out of the way, be easy to operate and shouldn't interfere with furniture.

I will describe here three types of porch windows. One is an old standby. The other two I've made and installed with the help of my inventive uncle, Dave Stewart.

All my window systems use ready-made sash either in stock sizes from a lumberyard or as custom-made units from a millwork shop (though you could certainly make your own, if you're so inclined). These systems are also based on the assumption that the porches won't be heated full-time. Porches designed for more than occasional cold-weather use should probably incorporate factory-made units because site-built windows are difficult to weatherstrip effectively.

Swing-up sash—Swing-up windows (drawing, top left), hinged to the head jamb, are traditional favorites for porches and basements. They aren't expensive, and when open, look appealing in an old-fashioned way. Installing them is fast and easy, particularly if the finished openings are precise and square. If they're not, you will have to use sash slightly oversize and trim them one at a time to fit. To install the sash, just space it about ¼ in. above the sill and screw hinges to the header and the top rail of the sash. Hooks and screw eyes hold the sash open, and turnbuttons screwed to the sill hold them closed.

A lot of older cottages in the lake country of Wisconsin and Minnesota use this window system. But when a summer thunderstorm blows up, you have to rush around and move chairs, lamps and maybe a stray knitting bag in order to swing the windows down and keep the rain out. High sills and low furniture minimize the problem, but of course that cuts into the view.

Accordion sash—Many years ago my grandfather devised an improvement on the old swing-up sash system. His system consists of sash, hinged to fold up like an accordion, that slide along an overhead track made out of garage-door hardware (photo previous page). The dimensions of the finished opening are not as critical for this system because the top of the sash butts against the side of the head jamb, and the custom hanger bracket allows a vertical adjustment (drawing, middle left). This system is well suited to retrofits and structures that are likely to shift a bit over the years. On my installation I had the hanger brackets made by a small machine shop, but I suspect you could make do with one leaf from a large butt hinge.

To install the system, just fasten the sash together with butt hinges and sit them on ¼-in. spacers on the sill. Slide the hanger rail onto the rollers, pull it up snug, and screw it to the header. If the window doesn't operate smoothly after you remove the spacers, it's easy to adjust the hanger brackets in place by tightening or loosening the nut at the bottom of the roller shaft.

The units I built run wall to wall, as do the tracks. Once opened and folded, the windows lie flat against the sidewall, where they are held in place by a wooden latch mounted on the wall above them (photo next page). When the win-

Drawings: Christopher Clapp; Photos: Douglass Ferrell

dow is closed, turnbuttons along the sides and bottom of the assembly secure the sash.

These windows are a little cumbersome to operate, partly because a large window is fairly heavy, although well-fitting windows up to about 14 ft. wide work fine after a little practice. Rolling one window unit toward each corner allows a full-width opening in a 28-ft. or 30-ft. wall—big enough for most houses. Windows with low sills can theoretically clear furniture placed along the wall by pulling and folding the sash from the corner while the unfolded sash roll toward you, but of course there is some tendency for the sash to fold up in the middle of the unit while you're pulling from the corner.

Sliding sash—My inspiration for sliding windows came from an old cottage whose windows had a groove in the bottom of the sash that slid on top of a rail on the sill. While this approach might work fine with smaller (and very straight) sash, swelling wood in summer humidity jammed nearly every window in the old cottage.

With the help of my brother and uncle, I devised a system of sliding windows that works much better and minimizes problems with low sills. In this system, the sash roll on wheels let into their bottom rails and are guided by parting stops screwed to the sill and head jamb (bottom drawing, facing page). The wheels were supplied installed in the sash by a millwork shop, but similar ones are available from large hardware outlets. I used nylon wheels instead of steel because they are quiet and won't rust.

This window system looks straightforward, but the installation can be time-consuming. The window-jamb dimensions are critical with this system since each sash slides along the entire opening and can't be trimmed to fit in specific places. On the installation I did, I wanted to minimize casing, so I framed the rough opening carefully, planning to fasten the sills and jambs directly to the framing. But I had to shim the jambs in some places anyway. Of course this meant that the sash for these openings had to be cut down. One off-size header, a slightly bowed post and a few other reminders that wood is a natural material convinced me that next time I would shim a finish frame into the rough opening just like a door jamb. This would ensure a precise opening for the sash to roll in without time-consuming trimming and adjusting.

The sash are held in place with stops and slide to one end of the opening. I used clear redwood for all the stops, jambs and sills because it is durable, easy to work, and resists warping. The bottom stops have a drain slot routed across their bottoms every 18 in. so they don't trap water on the sloping sill. They are screwed to the 6/4 redwood sills with brass screws. The sills themselves are screwed to the rough opening from below to minimize holes and hardware in the top of the finish sill.

On one 18-ft. wall where I used these windows, four 1⅜-in. wide sash roll on the sill, and I was concerned that the required sill would look awfully wide. Consequently I used narrow stops, which I wouldn't do again. I used ½-in. by ½-in. stops on the bottom and ½-in. by ¾-in. stops on the top. A better size would have been ¾ in. by ½ in. on the bottom and ¾ in. by 1¼ in. on the top. These larger stops would leave enough room between the head jamb and the top of the sash to pick the sash up and out of the track for cleaning and repair. Also, the stiffer track would better support the heavy sash.

The sill width I ended up with, 7¼ in., doesn't look cumbersome, given the big opening. Another inch or so in width, with a correspondingly wider stop, would have been better.

The sliding sash are sealed along their length with brush weatherstripping (bottom drawing, facing page). This operation went very smoothly. I worked from inside with the sash mounted in their tracks in the closed position. At each place where two sash came together, I fastened a strip of screen bead to the outer sash. Then I ripped 3/4 stock a little narrower than the sash for use as meeting stiles.

After cutting the meeting stiles to fit between the stops, I attached a piece of brush weatherstripping along their edges. This weatherstripping comes in rolls and has an adhesive backing (I added a few staples for insurance). The stiles were then each held in place against the screen bead, scribed and planed to fit flush with the face of the inner sash. With such big windows, this step was necessary because the stiles were not all perfectly straight, and the meeting stiles needed to fit snugly against the full length of the screen bead.

Once all the fitting was done, I screwed the meeting stiles to the sash with brass screws. If the sash warp in the future and ruin the fit of the weatherstripping, the meeting stiles can be unscrewed, adjusted and refastened. To make it easy to line up these weatherstripped joints when closing the windows, I screwed a stop block to the head jamb to locate each window.

Because wood does shrink and swell, the tolerances are tricky in any system where wood slides in wood. Of course a warped rail is also a source of trouble. When I visited the house two years after building my sliding windows, I could still operate each sash with one hand and quickly open or close a whole bank of windows. □

Douglass Ferrell lives in Trout Creek, Mont.

Accordion windows are held in place against the wall by a wooden latch.

Casing a Double-Hung Window

A job you can do quickly and well

by Bob Syvanen

Despite the growing use of casement windows in new construction, the double-hung wood window is still the traditional choice in Colonial-style houses. In the old days, double-hung windows became popular because they didn't require hinges or other hardware and because small panes of glass (often all that was available) could be used to make windows of different sizes.

Almost any salvage yard will have on hand a varied selection of old double-hung units for a fairly cheap price. The best new double-hung windows have insulated glass and a friction-type slide mechanism that doubles as a weather seal. Some are clad with vinyl. Whatever the brand or age, the anatomy is basically the same (drawing, right), and they're not difficult to case.

If the window has been installed right, its frame will be plumb and square. This makes casing a lot easier. Unlike doors, windows have frames that aren't usually flush with the surface of the wall. Most frames are a little proud, so you have to plane the top and side jambs flush with the wall before installing the trim.

Side and head jambs in a standard double-hung window are usually a full 4½ in. wide. They're designed for the thickness of a standard 2x4 wall: 3½-in. wide studs plus ½-in. drywall and ½ in. for exterior sheathing.

Window frames in thick walls will require extension jambs (drawing, below), which you can either order from the manufacturer or make yourself. These side and top pieces can be made from wood strips as thin as ⅛ in., or with 1x

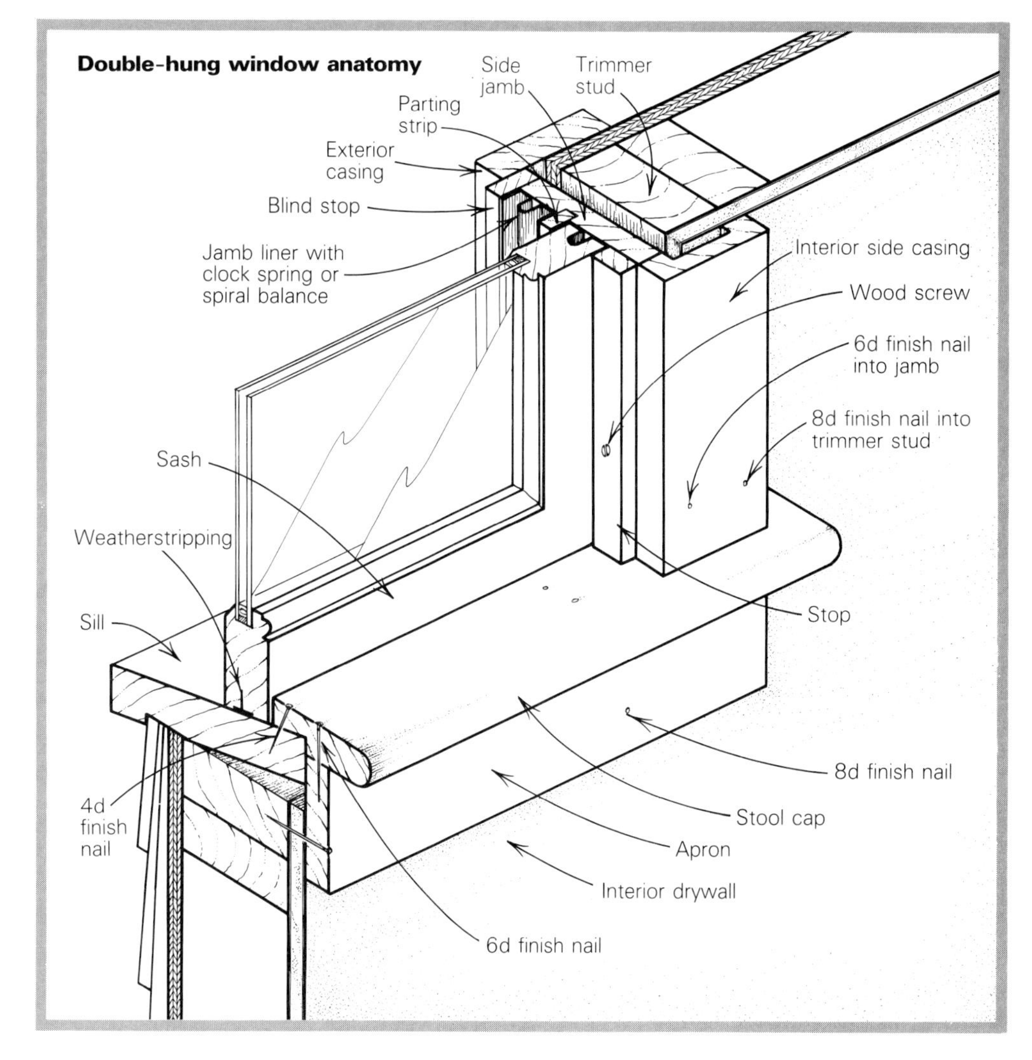

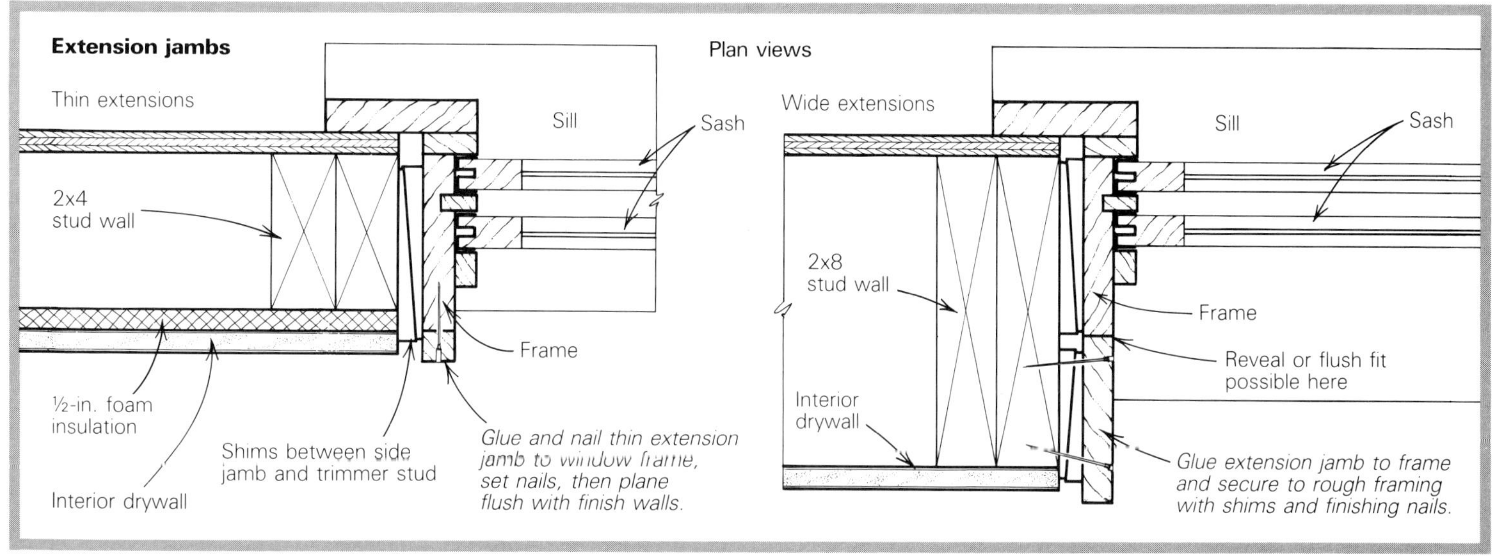

From *Fine Homebuilding* magazine (June 1986) 33:53-55

boards 6 in. or even 8 in. wide. The narrow-strip extension jambs can be glued and nailed directly to the edges of the existing window frames. If you use thin extension jambs, be sure to set the nails so that they won't nick your plane blade when you're planing the jambs flush with the interior wall surface. Wider extension jambs should be shimmed plumb and square with the window and nailed to the rough-opening studs and headers. I still make these a little oversize, so that they project past the surface of the wall; then I plane them in place to fit.

Installation—Many newer-style double-hung windows (Andersen and Pella, for example) can be cased on all four sides with the same trim stock. The stool cap and apron are optional. On older double-hung windows and in any Colonial-style house, you need to trim out the base of the window with a stool cap and apron. Most lumberyards stock about five basic stool-cap styles (drawing, below left). Older windows usually have a stool cap that's beveled on its underside to fit on top of the window sill, which is sloped to a standard 12° pitch. New windows use a stool cap that has a tongue on the inside edge that fits into a groove in the sill.

As bought from the lumberyard, the stool cap is just a pine plank that's been milled with a tongue or bevel. You have to cut it to fit the width of the sill and scribe the ends to fit against the wall, beneath the side casings.

To calculate the finished length of the stool cap, I use the length of the head casing and add 1½ in. to it. A stool cap of this length has nice lands at the bottom corners of the window, where side casings meet the stool cap (drawings, below right). To start, I cut the stool cap to around 2 in. longer than its finished length.

I center the stool cap on the sash, mark the cutlines for its fit over the sill and scribe the ends to fit snugly against the wall. When doing this layout work, allow for a slight gap between the cap and the sash so that the window will operate freely. If the wood is to be stained, a dime's thickness is good; use a nickel if the woodwork is to be painted.

Once the stool cap fits well against wall and sill, I cut the ends to their finished lengths. The ends have to be profiled to match the nosing at the front of the cap. I do this by removing most of the wood with a block plane. Final smoothing is done with sandpaper wrapped around a wood block. It's also possible to use a router if you have a bit that's close to the correct nosing profile. Also, I like to prime the underside of the stool cap where it sits on the window sill, since it's practically exterior trim, and I back-prime all exterior trim to curtail moisture absorption. Finally, I nail the stool cap in place with 4d finishing nails driven into the sill.

A variety of casing stock is available, so you're not limited to the square-edged 1x4s shown in the drawings. The side and head casings are fit in much the same way as you'd trim out a door ("Casing a Door," pp. 49-51). Some windows are

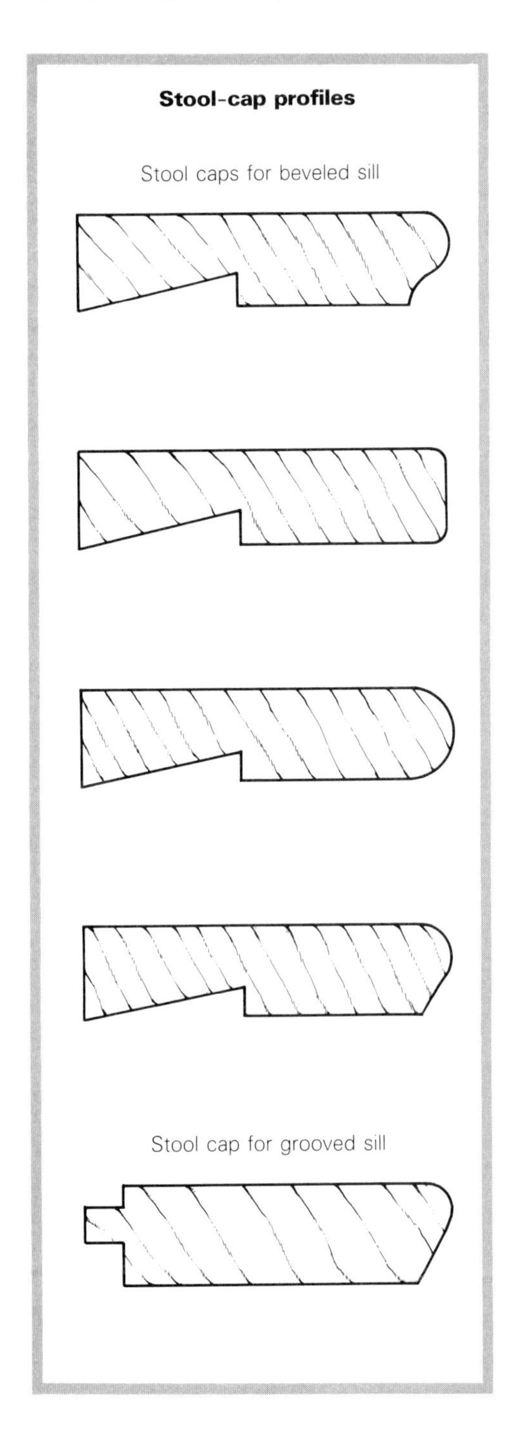

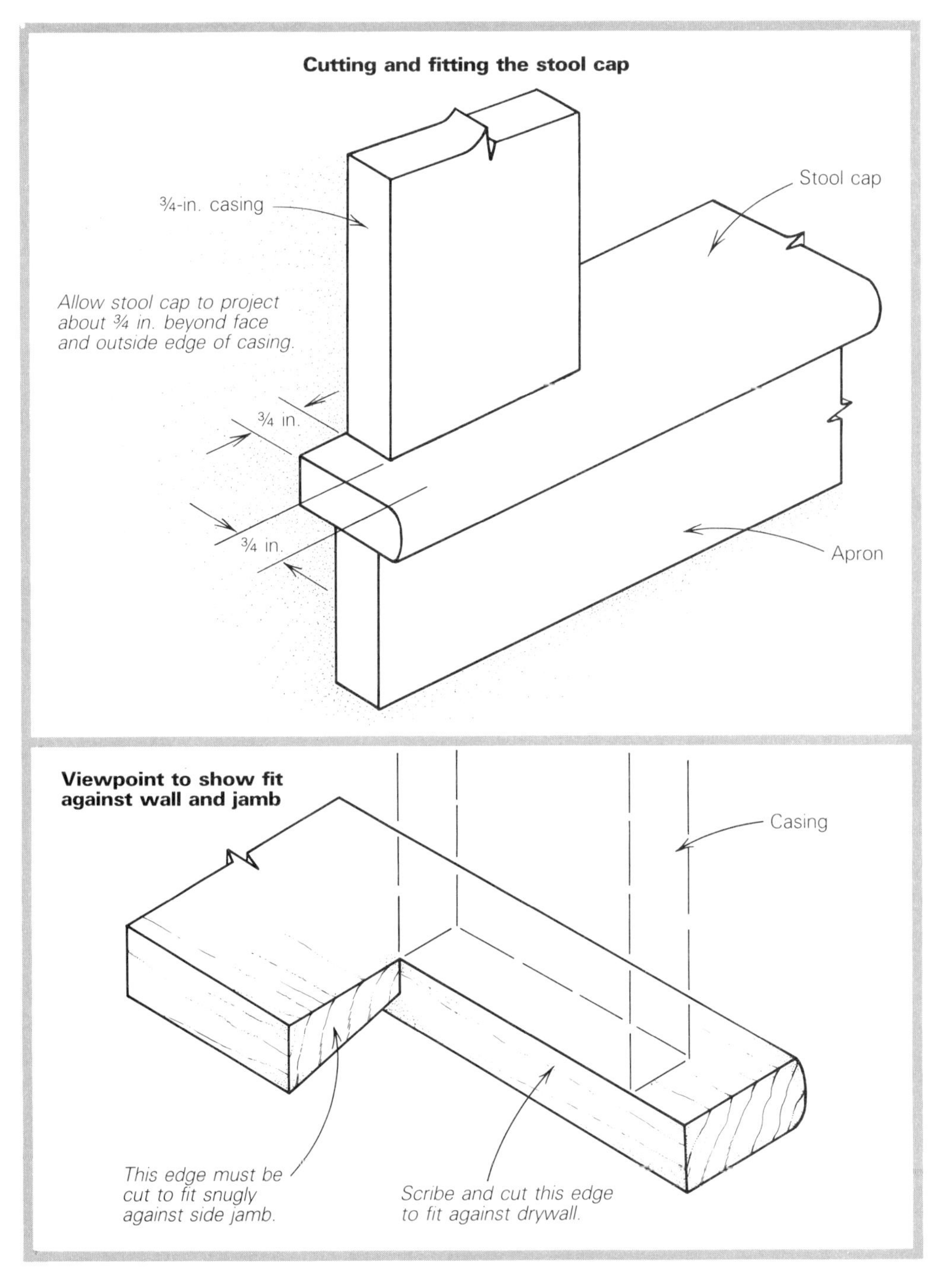

designed so there can be a slight reveal where the casings meet the jambs. Others (like the window shown in the drawings) have a flush fit.

The head casing can meet the side casings in a simple butt joint or in a miter, or all three trim pieces can butt into a pair of decorative corner blocks. As is the case with door trim, the ornateness of the trim depends on your taste, but you should always aim for tight-fitting joints. The pieces can be temporarily tacked in place to check for fit, then removed and trimmed with a chisel, block plane or chopsaw if necessary.

Once the joints are good, nail the casing in place with 8d finishing nails. If your trim pieces are less than ¾ in. thick, 6d or 4d nails will do. I nail the outer edge of the casing to the trimmer stud first, then nail the inner edge of the casing to the jamb. The trimmer connection is far stronger than the jamb connection, so I can straighten a slightly bowed casing, aligning it with the jamb without putting undue pressure on the jamb. I also drive a finishing nail up through the stool cap and into each side casing to snug the stool to the casings.

If you've installed a stool cap, the apron goes on next. It should be the same length as the head casing. If you're using molded trim, the apron will be cut from the same molding stock, and each end of the apron should be shaped by coping to match the profile of the molding. For a first-class job, especially if the trim is to be stained instead of painted, each end of the apron should have a mitered return instead of a cope (drawing, top right). The little return piece is very fragile and will split out if you try to nail it. So glue it instead, taking time to size the grain of both pieces with two applications of glue.

The apron has to be installed with the stool cap perpendicular to the window sash. A combination square placed against the window and the top of the stool is a good way to check this right angle (drawing, bottom right). If it's not right, a few raps with the hammer (use a scrap piece of wood to cushion the blows) should bring the stool cap up or down.

Before nailing the apron, I like to brace it from below with a short piece of scrap lumber that gets wedged between the apron's bottom edge and the floor. Then I nail down through the stool cap into the apron, and through the apron into the rough sill.

The last pieces of window trim to install are the stops, the small, molded pieces that run across the head and down each side of the jamb, defining the path of the innermost sash. On some double-hung windows, the stops are installed at the factory. If they're not, you should screw the side stops in place so that they can be taken off easily if you ever need to remove the sash. The top stop can be nailed with 4d or 6d finish nails. Again, if the window is to be painted, allow about 1/16 in. of free play between the edge of the stop and the window. If stain will be the finish, use slightly less clearance. Where stops join at the top corners of the window, I prefer to miter the molded part and butt the flat part. This creates a better joint than a full-width miter. □

Consulting editor Bob Syvanen lives in Brewster, Mass.

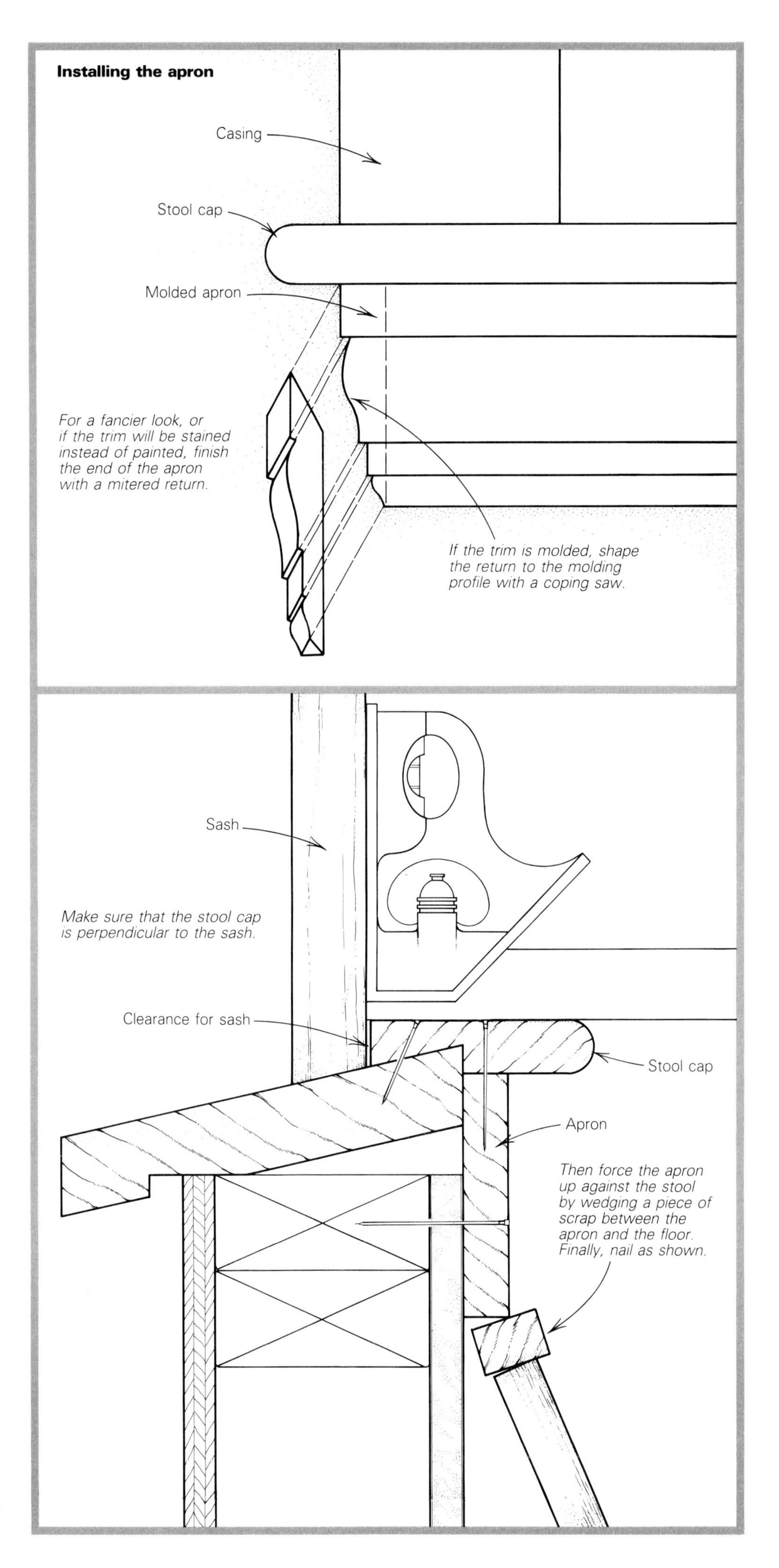

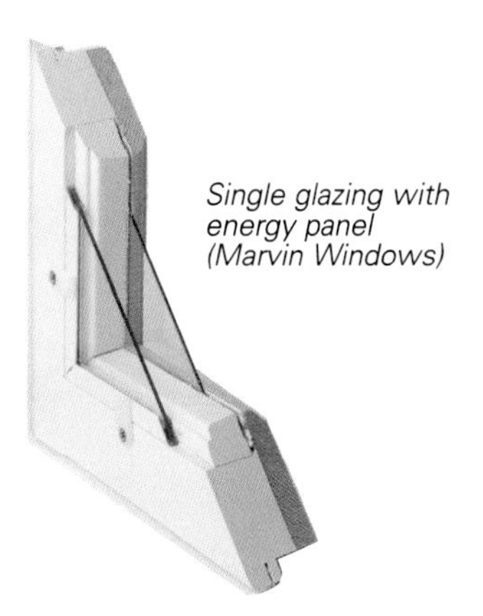
Single glazing with energy panel (Marvin Windows)

Double glazing (Marvin Windows)

High-Performance Glazing

Energy efficiency is the bottom line

by Paul Fisette

Twenty to thirty percent of an average home's heat is lost through its windows. But by using properly designed windows, you can improve your odds at saving energy. Multiple glazings, coated glass and gas fillings go a long way to keep heat in your home (or out, depending on climate) and money in your pocket.

There are many different glazing systems to choose from: single-pane, double-pane, triple-pane and quad-pane glazing systems (with or without removable energy panels); low-e coated glass; low-e coated plastic films; and argon gas-filled units (photos above). And many exotic new glazing ideas are being peeled from the drawing board (see sidebar, p. 85).

From outside to inside—When solar radiation strikes a building's surface, visible light, heat and ultraviolet radiation are either admitted into the building, absorbed by the building or reflected from it. Modern glazing systems aim to use or resist this energy drive.

Solar transmittance, the shading coefficient and the visible-light transmittance of glazing are important considerations when choosing a window. Each pertains to the passage of solar radiation of a specific wavelength through a window and into a building's interior.

Solar transmittance is a measure of how much total solar energy (visible, ultraviolet, and infrared) passes through glass. About 86% of the total amount of solar radiation that strikes ordinary single-pane float glass passes directly through it. The other 14% is either absorbed by the glass or is reflected by it. Additional layers of glass or coatings on the glass can dramatically reduce the amount of energy transmitted. Double-pane glazing transmits about 70% of the solar energy that strikes its surface. Triple-pane glazing transmits about 60%. Most low-e coated double-pane units transmit between 50% and 65% of the solar radiation, but versions that transmit even less are available (more on low-e coatings later). Visibility, daylighting and heat gain are directly linked to a window's solar-transmittance rating because this rating indicates the percentage of ultraviolet, visible and near-infrared energy transmitted through the glass.

Manufacturers use shading coefficients to describe how much infrared energy (heat) is transmitted by each of their glazing systems. The amount of heat that can pass through a single pane of ordinary float glass is the benchmark when it comes to quantifying a window's shading ability. A clear single pane of glass has a rating of "1." Other glazing systems are compared to this standard and are numerically rated according to how much heat they allow to pass through. For example, clear double-glazing allows about 84% as much heat to pass through as does clear single-pane glass, so its shading coefficient is 0.84. Double-pane low-e glazings typically score 0.65, but some are lower.

Visible light transmittance refers to the percentage of visible light striking a pane of glass that is transmitted through the glass. Today, residential windows can have high visible-light transmittance with either high or low shading coefficients. Manufacturers also adjust their window-coating schemes to suit each climate. In cooling climates, a window with a *low* shading coefficient works nicely, while in heating climates, a *high* shading coefficient works best.

Holding the heat in—How do we contain heat inside the living space once solar energy is transmitted through the glass? Here, absorptivity, reflectance and emissivity come into play. Every material can be rated according to its ability to absorb and emit (or radiate) energy. Absorptivity of a material equals its emissivity, so a good absorber is also a good emitter. A poor absorber is a good reflector.

Materials are assigned a numerical rating between 0 and 1 depending on their ability to absorb and emit energy of a specific wavelength. A material that absorbs and emits no energy is assigned a rating of 0. Aluminum foil scores about 0.1. A perfect emitter, which is called a blackbody, rates a 1. The value for a clear single pane of float glass is about 0.88, while the best low-e (low-emissivity) coatings typically rate about 0.10 to 0.15.

Low-e—The first low-e coatings were siphoned from research vats during the energy crunch of the 1970s. In the early 1980s, Southwall Technologies of Palo Alto, California, introduced Heat Mirror, a 2-mil low-e coated polyester film that is stretched within the airspace between the panes of glass. Now every major window manufacturer offers some sort of low-e window (see chart, facing page). And outfits such as Alpen, Inc. (5400 Spine Rd., Boulder, Colo. 80301) offer low-e glazing units that can be installed in shop-built sash.

From *Fine Homebuilding* magazine (August 1989) 55:78-81

Triple glazing (Pozzi Wood Windows)

Triple glazing with Heat Mirror (Hurd Millwork Co.)

Quad glazing with double Heat Mirror filling (Alpen, Inc.)

Low-e coatings reflect heat, while allowing the passage of visible light. They also reflect fabric-damaging ultraviolet rays. Low-e coatings require an airspace to work effectively, so they're used with multiple-pane windows (two or more panes of glass with airspaces between them).

Manufacturers offer low-e windows with either coated glass or with a coated plastic film (like Heat Mirror) suspended between panes of glass. The beauty of low-e coatings is that they reflect exterior heat away from the house in the summertime and hold heat in during winter (in a hot climate, they keep heat from entering year around). Low-e coatings allow short-wave solar energy to pass through the glass into the living space, where it is absorbed by floors, walls and household furnishings. The energy is transformed and reradiated from these interior surfaces as long-wave (heat) energy. When the long-wave energy tries to escape through the low-e coated glass, about 90% of the reradiated heat is blocked.

When coated glass is used in heating climates, low-e coating is applied to the *interior* pane of glass on the side facing the interpane airspace. This reduces the drafty feeling in a room because the interior surface of the glass is kept warm by the low-e coating. For cooling climates, the coating is applied to the *exterior* pane of glass on the side facing the airspace.

Low-e coatings are applied to a glass surface as either a hard or soft coat. Hard-coat glass is obtained by spraying metallic oxides onto hot glass. The oxides fuse to the molten surface of the glass. For soft-coat glass, silver and metallic oxides are "vacuum-sputtered" onto the surface of the glass. The silver acts as a space blanket—it reflects heat. Additional metallic coatings are applied over the silver to remove its ability to reflect light, making the finished product more aesthetically pleasing. Both hard and soft coatings are only molecules thick and are transparent.

Hard-coat glass is more durable and easier to manufacture than soft-coat glass, but its energy performance isn't quite as good as its soft-coat rival. Soft-coat windows are generally 20% more thermally resistant than hard-coat windows (R-3.1 vs. R-2.6), but several months' exposure to air will degrade soft coatings.

Window Manufacturer	Types	Frame material	Glazing	R-value	Shading coefficient
Andersen Corp. 100 4th Ave. North Bayport, Minn. 55003	C, A, DH F, S	W, VCW	DG TG LE LEA avail.	2.00 3.10 3.2	.89 to .34
BiltBest Windows 175 Tenth St. Ste. Genevieve, Mo. 63670	C, A, DH F	W, ACW	DG TG LE	2.35 3.10 3.30	.89 to .74
Eagle Windows and Doors, Inc. 375 East 9th St. Dubuque, Iowa 52001	C, A, DH F	ACW	DG LE LEA	2.04 3.13 3.85	.92 to .68
Hurd Millwork Co. 520 S. Whelon Ave. P.O. Box 319 Medford, Wis. 54451	C, A, DH F, S	W, ACW	DG LET Heat Mirror 88)	2.3 4.05	.88 to .41
Kolbe & Kolbe Millwork Co., Inc. 1323 S. Eleventh Ave. Wausau, Wis. 54401	C, A, DH F, S	W, ACW	DG LE LEA avail.	1.92 3.29	N/A
Marvin Windows Hwy. 11, P.O. Box 100 Warroad, Minn. 56763	C, T, A DH, F, S	W, ACW	DG LE LEA	2.27 3.45 4.35	N/A
Norco Windows, Inc. 811 Factory St. P.O. Box 140 Hawkins, Wis. 54530	C, A, DH F, S	W, ACW VCW	DG LEA	1.96 3.33 to 4.35	0.91 to 0.35
Openings, Inc. Hobeka Windows 14 Vernon St., Suite 201 Framingham, Mass 01701	C, T, F S	W	DG TG avail LE LEA	2.5 3.5 4.5	.89 to .34
Paeco, Inc. One Executive Dr. P.O. Box 968 Toms River, N. J. 08753	C, A, DH	W, ACW	DG LE	2.05 2.75	N/A
Peachtree Doors, Inc. P.O. Box 5700 Norcross, Ga. 30091	C, A, DH F	AE/WI	DG LE	2.44 3.45	.89 to .64
Pella Windows Rolscreen Company 100 Main St. Pella, Iowa 50219	C, A, DH F	W, ACW	DG LE LE (shaded) LEA avail.	2.13 3.03 4.3	.88 to .19
Pozzi Wood Windows P.O. Box 5249 Bend, Ore. 97708	C, A, DH F	W, ACW	DG TG LE	2.33 2.94 2.86	.89 to .34
Traco Cranberry Industrial Park P.O. Box 805 Warrendale, Pa. 15095	C, A, DH F, T, S	V, A	DG LE	1.85 2.35	N/A
Weather Shield Mfg., Inc. 531 Northeast St. P.O. Box 309 Medford, Wis. 54451	C, A, DH F, T, S	W, VCW ACW	DG TG LE LET	2.37 3.03 3.00 3.5	N/A

Types: C = Casements; T = Tilt; A = Awning; DH = Double-Hung; F = Fixed; S = Sliding

Glazing: DG = Double-Glazed; TG = Triple-Glazed; LE = Low-e Double-Glazed; LET = Low-e Triple-Glazed; LEA = Low-e with argon gas filling.

Frame material: W = Wood; ACW = Aluminum-Clad Wood; V = Vinyl, VCW = Vinyl-Clad Wood, AE/WI = Aluminum exterior with wood interior.

Note: Some of the manufacturers listed have argon gas-filled units but did not have tested values available. Ask your dealer about this option.

Edge-seal details

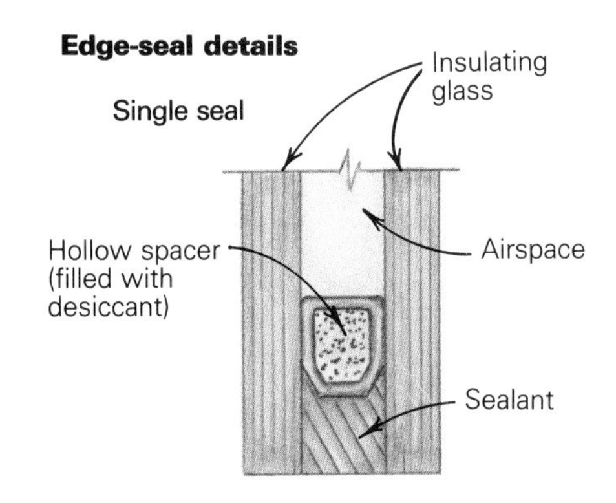

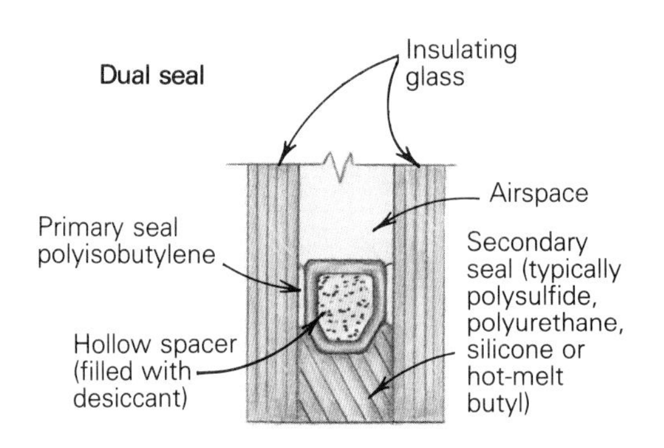

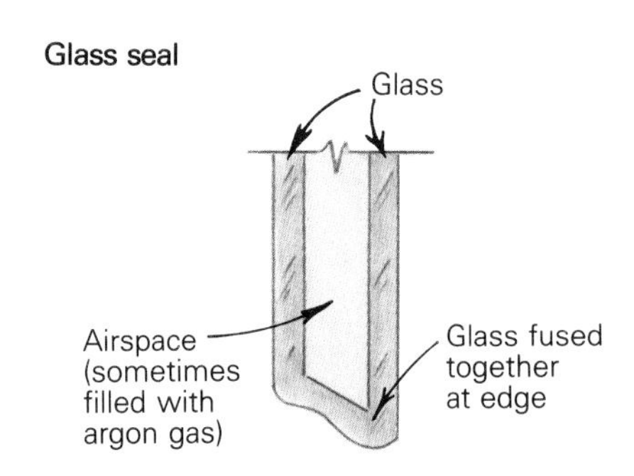

Therefore, soft-coat windows are only as good as a window's edge seal.

Some low-e coatings show a faint blue tint, and under some lighting conditions they will exhibit a perceptible iridescence. But for the most part, low-e coatings are difficult to detect.

R-values—For a window to be considered energy-efficient, it must block the loss (or gain in a cooling climate) of heat energy. Air does a fairly good job of blocking heat loss. Trapping air between panes of glass boosts the thermal resistance of any window.

The chart below compares R-values of various glazing systems. Typically, triple-pane and quad-pane units are heavy and expensive (triple-glazed units weigh 50% more than similar-size double-glazed low-e units). Low-e units cost less than triple-glazed units and perform better to boot. Some manufacturers stretch low-e coated plastic film within the airspace of double-glazed units to provide an effective third or fourth "pane." The weight of these units is similar to conventional double-glazing, but performance mimics that of triple-pane or quad-pane units.

Argon versus air—The use of argon gas in place of air in multiple-glazed windows is catching on in the U. S. (argon has been used for this purpose in Europe for years). Argon is heavier and denser than air, so it boosts the R-value of a window and helps to deaden sound. Also, argon is most effective at a narrower gap size than air, allowing manufacturers to use skinnier windows.

Many window manufacturers are offering argon in combination with low-e glass (in fact, argon is effective only when it is used with low-e glass). Some manufacturers claim that if the airspace between low-e coated, double-pane glass is filled with argon gas, the window's thermal rating jumps to R-5. But John Schlueter, technical services engineer with PPG, the world's largest producer of glass, says, "I would be very surprised and very interested in seeing an argon-filled double-glazed window with an R-5 rating." He contends that a low-e double-glazed window provides a base thermal resistance of R-3, and that when argon gas fills the space instead of air, the R-value jumps to around R-4.

If the window's edge seal is not perfectly tight, the argon gas will leak out and the high R-value associated with it will be thrown, well, out the window. Argon gas can eventually leak through even a well-made seal, leaving the window only as good as a regular low-e window.

Schlueter is a member of the American Society for Testing and Materials (ASTM) committee that is currently developing methods for testing the loss of argon through a glazing's edge seal. According to him, the ASTM committee has yet to establish what is a reasonable loss. "The consideration becomes: how much do you pay for argon?" Some window manufacturers claim they are filling their units with argon at no extra charge to the consumer. Others promise long service life by offering a 20-year guarantee against leakage. And PPG is the only glass manufacturer that makes glass-sealed edges for double-pane glass. The prospect of glass-sealed edges combined with low-e coated glass and argon filling implies permanent argon filling.

On the edge—It is virtually impossible to construct a window that does not have a thermal bridge of some type. In a double-pane window, for instance, the two panes of glass are separated by an airspace, but the edges are connected. Heat loss is always greatest around the edge of the glazing because of this bridge. Check out your windows on the next cold morning. You'll see that condensation accumulates at the edges of the glazing first because the glass is coldest here.

The choice of material used to seal the edges of the glass can substantially affect the rate at which heat is lost. There are three basic types of edge seals: single-seal, dual-seal and glass-seal (drawings above). With single-seal and dual-seal systems, the panes of glass are separated by a hollow spacer (usually metal) filled with a desiccant that absorbs any moisture left between panes of glass during the manufacturing process. This helps prevent condensation from forming within the airspace. Only one sealant is applied (to the outboard edge of the spacer) in a single-seal system. This is the least-durable seal.

A dual-seal system is better. Here, a polyisobutylene sealant is applied between the panes of glass and the spacer. Then a second seal—usually polysulfide, polyurethane, silicone or hot-melt butyl—is applied along the outboard edge of the spacer. The primary seal blocks moisture and air migration, so condensation doesn't collect between panes of glass. The secondary seal assures the structural integrity of the unit. The seal on a dual-seal unit is less likely to fail than on a single-seal unit as a result of the material's deformation during swings in barometric pressure and temperature. A properly constructed dual-seal unit will easily last 20 years. This seal, unlike single seals or glass seals, thermally uncouples the inner and outer surfaces of the window, cutting heat loss.

With glass seals, the two panes of glass in a double-pane window are fused together along the

Efficiency and Cost

Glazing system	Approximate R-value	Approximate relative cost
Single glazing	1	.9
Double glazing	1-5-2.0	1
Triple glazing	2-5-3.0	1.15
Low-e double glazing	3.0	1.1
Low-e double glazing with argon fill	4.5	1.1
Double glazing with low-e plastic interpane	4.5	1.2
Double glazing with low-e plastic interpane and argon fill	5.0	1.25

R-values vary according to the type of glass, the low-e coatings used and the thickness of the interpane air space (½ in. to ¾ in. thick is optimum). R-values based on manufacturer's literature.

Condensation

Glazing system	Predicted temperature of inboard pane of glass	Indoor relative humidity required for condensation to appear
Single glazing (⅛-in. glass)	20°	15%
Double glazing with ½-in. air space	47°	45%
Low-e double glazing with ½-in. air space	52°	53%
Triple glazing with ½-in. air space	53°	55%
Low-e triple glazing with ½-in. air space	54°	57%

Based on an indoor temperature of 70° and an outdoor temperature of 0° with a 15 mph wind.

Drawing: Michael Mandarano

edges. As mentioned earlier, glass seals hold promise for containing argon. Unfortunately, they also make good thermal bridges.

Cold sweat—In addition to saving heat, energy-efficient windows also reduce or eliminate condensation problems. Condensation is the fallout in the war between temperature and relative humidity (bottom chart, facing page). Plain and simple, warm air can hold more moisture than cold air can. Condensation will collect on a surface that is colder than the dew-point temperature of the air it is exposed to. For example, when the indoor air temperature is 70° F at 50% relative humidity (common indoor wintertime conditions), moisture will collect on a 50° F surface. And often this surface is the glass in your window. Warm glass prevents condensation from forming, so energy-efficient windows are the cure.

What about condensation that forms *between* panes of glass? Joe Klems, staff scientist at Lawrence Berkeley Laboratory in California, thinks this condensation can be more than just a nuisance. Klem says: "Soft coats will degrade when they are exposed to humid air." Even if the low-e coating doesn't degrade, it will be deemed useless because water has a high emissivity. Condensation cancels the value added by the coating.

Poorly constructed edge seals allow moisture to collect between the panes, but Klems points to another cause of leaky edge seals: "If sealed units are shipped over mountains, the seals can be damaged by the air-pressure change. Also, problems can occur when windows are installed at an altitude different from where they were manufactured. A 1,000-ft. change in altitude can be enough to break a seal." But many glass manufacturers argue that Klems is overly cautious in his assessment, saying that it takes a change of 4,000 to 5,000 feet in elevation to affect an edge seal.

What are manufacturers doing about this problem? Some, such as Marvin Windows, incorporate capillary tubes made of aluminum or stainless steel into the edge seals. The tubes allow the pressure inside a glazing unit to equalize with the pressure outside. Marvin Windows claims that these capillary tubes also prevent moisture from entering the airspace. Some manufacturers are fitting their glazing units with breather tubes instead of capillary tubes. These breather tubes are crimped off once the windows reach their destination. But because breather tubes are accessible only before the glazing unit is installed in its window frame, that destination is usually a window manufacturer's plant and not the job site itself. If the job site is located at a different elevation than the plant, breather tubes won't solve the problem. It appears as if there's more work to be done before the altitude problem is solved satisfactorily. □

Paul Fisette is a visiting professor and coordinator of the Building Materials Technology and Management Program at the University of Massachusetts at Amherst.

Windows into the future

by Kip Park

For years, windows have been the weakest link in the energy systems of houses. But thanks to emerging "Buck Rogers" technical developments, windows may soon provide as much insulation as the walls they're mounted in. A recent one-day seminar held by the Institute for Research in Construction of the National Research Council of Canada—and the presentation by Steve Cornick (an IRC researcher based in Calgary, Alberta) in particular—offered a hint of things to come. Though the conference was specifically attuned to Canadian research and development, the U. S. appears to be on a similar track.

***Ventilator windows*—One intriguing new technical development is the ventilator window. Ventilator windows, which are triple-glazed, draw ventilation air into the home through the space between the two outer panes of glass. The windows rescue heat that would otherwise be lost from the interior of a building through convection, conduction and radiation, and they preheat ventilation air in the process.**

One version is being developed by Willmar Window Industries Ltd. (485 Watt Street, Box 99, Station F, Winnipeg, Manitoba, R2L 2A5) in connection with Dr. Gren Yuill, a Winnipeg consulting engineer. Willmar and Yuill were awarded two research contracts worth $100,000 from Energy, Mines and Resources Canada to develop, test and evaluate PVC-framed prototype ventilator windows. Their windows were installed in a demonstration house in the fall of 1988. The house is designed to operate under negative pressure, and the windows' air inlets are controlled by units that sense humidity in a room. The inlets are always open (they each admit from 6 to 10 cfm), but the opening varies according to relative humidity in the room. The inlets, called "humidity-controlled air-inlet ventilation systems," are made by American ALDES Ventilation Corp. (4539 Northgate Ct., Sarasota, Fl. 34234). Computer predictions of nighttime performance rate this window at R-6—double the R-value of a typical R-3 rated triple-pane window.

***Switchable glazings*—Switchable glazings also hold some promise. One type uses an electrochromatic material like that found in the liquid-crystal displays of calculators. The liquid crystals are part of a thin film that is sandwiched between two panes of glass. When an electric current is passed through the film, the film becomes clear. When the current is switched off, the film turns a silvery, metallic grey that reflects heat while offering a measure of privacy. In fact, these windows could make curtains obsolete. Willmar Windows is working on this technology, too, and a liquid-crystal switchable window may be available commercially within two years. In the U. S., Marvin Windows is running similar experiments with an eye toward the marketplace.**

Another type of switchable glazing uses thermochromatic materials, which alter a window's insulative and optical properties in response to changes in temperature. These materials are usually gels or liquids sandwiched between two panes of glass. Drawbacks include reduced transmission of visible light and a possibility of leakage of the liquids through the window's edge seals.

Still another version of switchable glazing makes use of photochromic materials, which are sensitive to light. When struck by sunlight, these materials, like those used in photo-grey sunglasses, reduce their visible-light transmittance and their shading coefficient (they let in less light and heat).

***Aerogel windows*—Another innovative window uses aerogels, sometimes described as transparent insulation, between layers of glass, giving between R-5 and R-7 per inch. The silica gels consist of a bonded network of silica pebbles, each smaller than the wavelength of visible light, which means that visible-light transmittance is unaffected by them. But, because the space between the molecules is too small for air molecules to pass through, aerogels cut down on convective and conductive heat loss. There are technical problems with aerogels, though. First, they are fragile, which makes them difficult to support in a window. Second, the manufacturing process is dangerous, requiring a high-pressure autoclave to produce the aerogels.**

***Evacuated glazing*—Manufacturers are also looking at evacuated glazing, which gives better thermal performance than standard sealed, double-glazed windows do. An evacuated unit consists of two panes of glass just thousandths of an inch apart with a low-e coating on one of the panes and as much air as possible removed from the interpane space. This produces a window with an insulating value of about R-10. With the low-e coating, the window is like a thermos bottle.**

Again, there are problems with evacuated glazing, especially with sealing the edges. Currently, lasers are used to edge-seal the glass. Another problem is in maintaining the distance between the panes of glass, especially under wind load. Because the largest window containing evacuated glass is only about 10 sq. ft. in size, Cornick thinks this kind of window is "down the road a bit yet."

In addition to these futuristic glazing schemes, advances such as PVC and pultruded fiberglass window frames, insulating spacers to increase temperatures around the edges of the glass (by providing a thermal break), new types of weatherstripping and the use of computer-controlled machinery in window manufacturing plants will all contribute dramatically to improved energy performance of windows.

Kip Park lives in Winnipeg and writes about housing, construction and energy technology.

Installing Fixed Glass Windows

Double-glazed units that don't move let in light and heat, but keep out drafts

by Dale McCormick

Windows have traditionally provided light and ventilation, but asking them to perform both functions isn't always wise, especially when many of us are installing large expanses of glass on our south walls. Windows that don't open can be sealed more tightly than those that do, and they cut down on the infiltration that often amounts to 40% of a house's heat loss. Ventilation can be handled by screened wall openings, carefully placed and built to be heavily insulated and tightly sealed during cold weather. Even if you like movable sash for ventilation, you can intersperse it with fixed panes.

Insulated glass—A single pane of glass is a lousy insulator, typically yielding an R-value of less than 1. Two spaced panes trap a layer of still air between them and can cut heat loss in half. Three panes create two air spaces and cut heat loss even further, but triple glazing isn't cost-effective in most areas, especially if you're planning to use some form of insulation inside the windows at night. The problem with double glazing has always been to eliminate condensation between the panes while at the same time sealing them to prevent convection from destroying the insulation value of still air. Although you can build your own double-pane insulating windows by installing separate sheets of glass in a wood frame, moisture will invariably migrate through the wood, and the resulting condensation looks bad and can lead to rotting frames. Weep holes aren't a good solution—you don't want air movement. I recommend buying ready-made double-glazed units.

Commercial insulated windows are built of two panes of glass separated in an aluminum channel. The channel contains silica gel, a dessicant that absorbs moisture trapped between the panes of glass at the factory. The glass and channel are sealed with either one material or two (drawing, above). When only one seal is used, it is usually hot-melt butyl or one of the varieties of polysulfide. Such units are often guaranteed for five years. Better windows have two seals. The primary sealant, a moisture barrier, is polyisobutylene. The edge sealant is either polysulfide or silicone. Don't confuse any of these materials with the glazing sealant you or your contractor will have to apply when installing the glass. It's important for the factory edge sealant and your site-applied glazing sealant to be compatible—many aren't.

Glass
Dead air space
Dessicant (silica gel)
Primary seal (polyisobutylene)
Edge seal (polysulfide or hot-melt butyl)
Single seal
Dual seal
Edge seal (polysulfide or silicone)

Factory-sealed insulating glass
Dual-seal units are more resistant to punctures or other damage that could destroy the insulating value of the air space. They generally cost more and come with longer guarantees.

Jambs—A jamb is a frame that holds the window and is set within the rough opening in the side of the house. Fixed-pane windows can be built either with or without a jamb. Using one means more work, but jambs are independent of the house's framing, and can be plumbed and leveled within the rough opening. If you don't use a jamb and attach your windows directly to the studs, the fate of the glass is wedded to the future of the frame. The key to a good marriage is dry wood and accurate framing. Generally, fixed windows with jambs look more finished, but you can also achieve an attractive effect without using a jamb, if you choose a handsome wood as the structural material for the window wall.

For large expanses of glass, it usually isn't practical to invest the time and material necessary to build a jamb for each pane. There are a number of ways to install double-glazed windows without jambs. One method we often use is shown in the drawing at the top of the facing page. The glass, attached to the outside of the studs, is held in place by the casing, which doubles as a stop. We set each glass unit to bear ⅜ in. on the studs, leaving a ¾-in. gap between units if we're framing with nominal 2x4s.

Manufacturers of insulating glass recommend that the units be installed with ⅛-in. interior and exterior face clearance between the glass and the stop. For this, we use butyl glazing tape all around, which functions both as a bed for the glass and as a dam to prevent the caulking from touching the edge seal. Butyl also reacts with the primary sealant, so keep their edges separated. This is easy if you use narrow tape.

We set the glass on two neoprene blocks measuring ½ in. by 4 in. by the thickness of the unit. These should be positioned in from the edge of the window by one-quarter the length of the glass panes. Last, we caulk with a waterproof material compatible with the edge seal-

Tilt? **While you are designing your house or addition, you have to decide whether to install your windows vertically or angle them to accept more solar radiation. I've stopped using slanted south-facing glass except in a few special situations: a space that will be used mostly to grow plants; a space that can be closed off from the house so that overheating in summer and night insulation in winter are not problems, or a retrofit of a narrow south porch, where slanting the glass from the roof eave to the ground will yield more room.**

Angled glass creates what is basically a glass roof, inviting a multitude of problems: leakage, breakage, overheating in summer and snow cover in winter. Also, in the summer, when you don't need it, many more Btus come through glass angled at 60° than through vertical windows—and vertical glass can be more easily shaded. On the other hand, there is a very small difference in performance between 60° glass and 90° glass during December and January, because the low winter sun is more nearly perpendicular to vertical windows then.

If you do choose to angle your glass, it's probably best to use a commercial-style aluminum glazing system with EPDM rubber gaskets (drawing, left). Aluminum won't expand or twist as wood does under the extreme conditions faced by a roof surface oriented south.

An aluminum glazing system

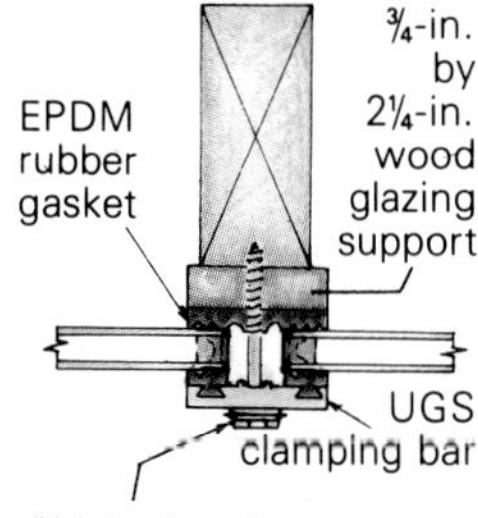

Illustrations: Barbara Smolover

From *Fine Homebuilding* magazine (April 1982) 8:42-43

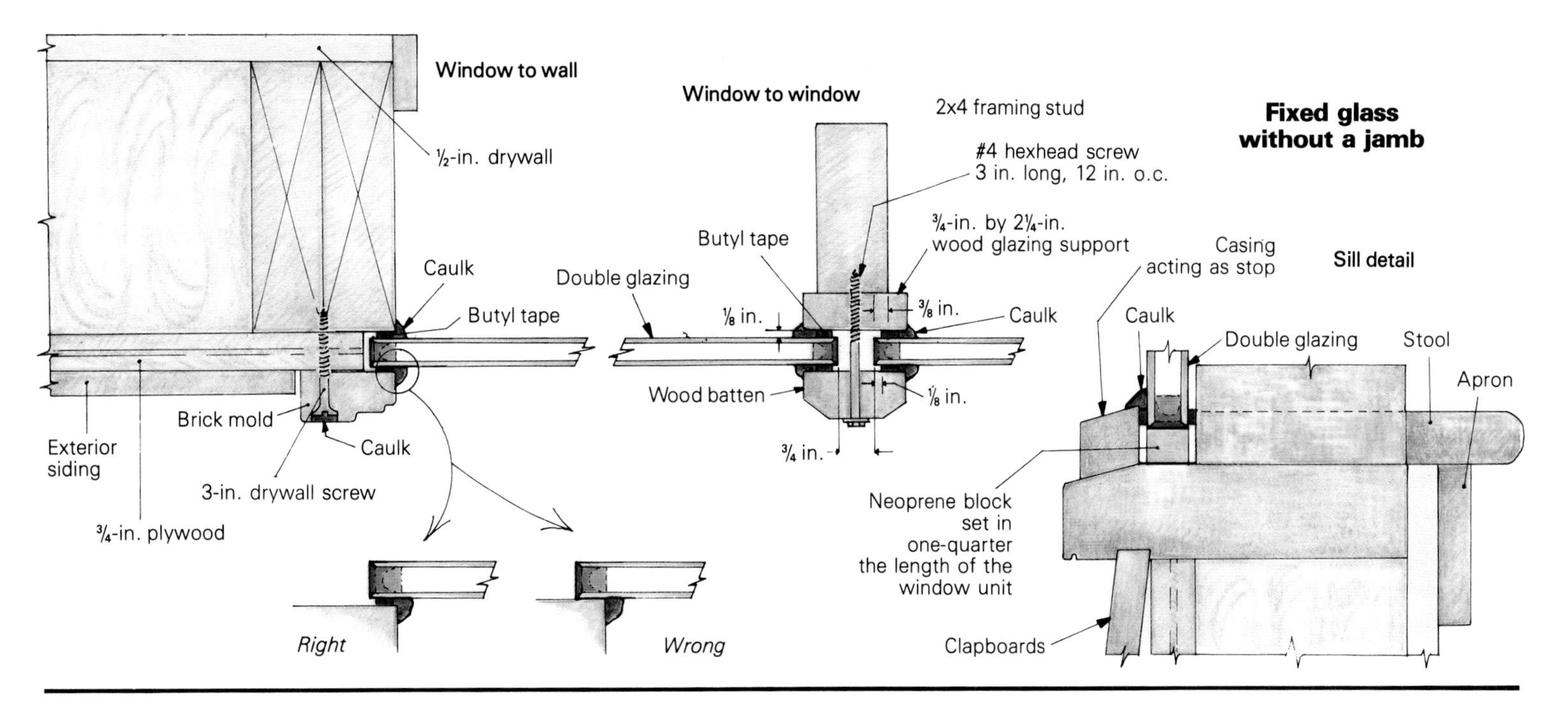

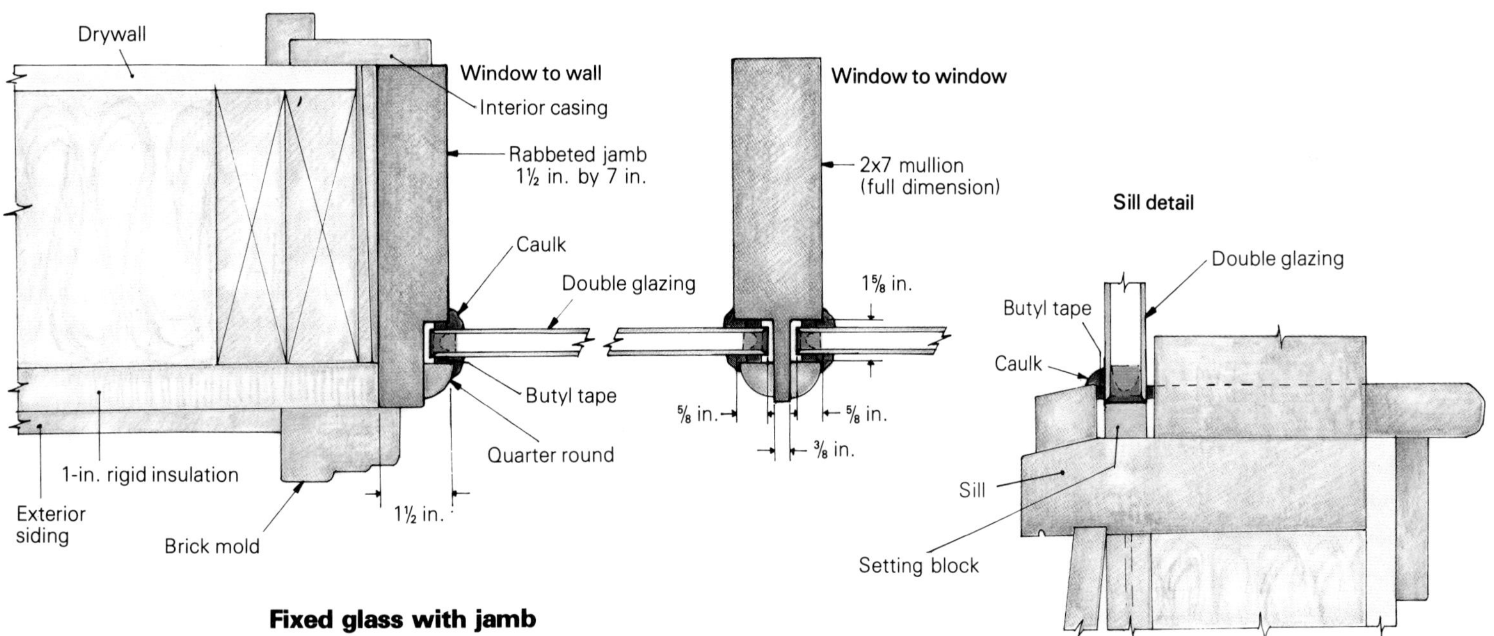

Fixed glass with jamb

ant. As shown in the drawing, caulk should actually cover an opening, not just be run along a crack. Wood should be primed before assembly, because many caulks, including silicone, pull away from bare wood over time.

Last summer, we built an inexpensive greenhouse this way, using pressure-treated lumber to withstand condensation and moisture. The greenhouse looks appropriately plain and practical. We also used this method of glazing on a sunspace kitchen/family room in a summer house we were retrofitting for year-round use. Here we used lauan studs to frame the south wall, and lauan exterior stops to hold the glass. The result is elegant. It's not the method but the materials (and the care with which you use them) that determine how your window installation will look.

For relatively small areas of fixed glass, especially those that are integrated with operable windows, I think it's best to design fixed windows with a traditional jamb that can be made in the shop. Exterior casing that matches the casing of movable sash will visually tie the two types of window together.

Jamb stock should be dry. Pine or Douglas fir is fine. Redwood is a good choice for greenhouses. Rip your stock to a width equal to the distance between the interior and exterior finish materials. You can rabbet your jambs, as shown above, to produce an integral stop. If you do, the rabbet depth should match the size of the stops that you're going to apply to the outside—usually ¾ in. The width of the rabbet should be at least 1 in. greater than the thickness of the glass, depending on the thickness of the glazing tape you'll be using. An alternative to the rabbeted jamb is to use 5/4 stock (which is really 1 in. thick) for the jamb and add 4/4 interior stops, which can also act as stops for insulated shutters.

Either way, it's best to build the window sill from 7/4 stock. As shown in the drawing, the sill is wider than the rest of the jamb, and should be designed to overhang the exterior siding by at least 1 in. The sill should be beveled so that rain will run off, and it should have a drip kerf to prevent water from running in along its bottom surface. A groove under the sill to accept siding material makes for an extra-tight installation.

In the shop, rabbet the tops and bottoms of the side jambs to accept the head and sill, keeping everything square. Glue and nail them into place with 12d ring-shank nails. The finished jamb unit can then be plumbed and leveled in the rough opening with shims under the sill and along the side jambs. When you install the glass, use the same clearances and techniques you would in a jambless installation, and be sure to leave ⅛ in. between the inside edge of the butyl tape and the interior edge of the rabbet. Brick mold is a good exterior trim treatment, because it closely matches the trim on commercially manufactured windows. □

Dale McCormick is in charge of building services at Cornerstones, an owner-builder school in Brunswick, Maine.

Leaded Glass

The tools, materials and techniques of this craft haven't changed much in 500 years

by Doug Hechter

It wasn't easy to find a large sheet of glass in the Middle Ages, but you could get small, bubble-filled glass discs called crowns. To make them, a glassmaker would gather a blob of molten glass on the end of a blowpipe, and spin it until it formed a circular plate, sometimes as big as 5 ft. in diameter. The resulting glass might be colored or clear, but it was always very expensive.

Once it hardened, the glass was ready to be cut into pieces and assembled by artisans into windows. These craftsmen used lead channels called cames to tie the pieces of glass together into a large panel.

Making windows this way was an expensive, time-consuming process, and the Church was just about the only customer who could afford them. Church windows often depicted Biblical events, thereby teaching the illiterate as well as letting in the light.

The beginning of the 15th century brought change. Church windows were becoming wider and taller, and a burgeoning middle class wanted more comfortable, better lit houses. Simple rectilinear windows, using square and diamond-shaped panes linked by lead cames, were made to satisfy both markets. These are called quarry windows (after the individual panes of glass, which are called quarrels), and building one is the topic of this article. Quarry windows remained the standard glazing for centuries, and there is a revived interest in them today. I think it's a reaction to the textureless modern goods that we see everywhere. The individual panes in a leaded-glass window are almost in the same plane, but not quite. The tiny differences in angle between them fracture the reflected light. This random quality is the human touch, a valued commodity in the late 20th century.

Designing the window—A good leaded window is a window first, and an object of art second. A sound window must withstand a reasonable wind load, as well as the sometimes grueling fluctuations of temperature and movement within the building. To achieve this, steel-reinforcement braces are soldered to the inside of the lead frame. Without them the window would eventually sag and fall apart. A well-designed leaded window has lead lines wide enough to hide the braces.

When a building is designed, the architect gives considerable thought to its proportions, its visual mass and the textures and values of the materials. The leaded-glass maker should tune in to these choices. A successful window design reflects the scale of the structure, and if it has colored glass, it should have light and dark areas, just as the building's mass presents highlights and shadows. Like the glazing bars in French doors, the lead lines afford some patterned continuity in the plane of the wall. The larger the window is, the heavier the lines should be.

Strong outdoor light tends to minimize the lead lines once the window is in place. It's especially noticeable in skylights. Colors that might look strong and expressive on the workbench can easily become anemic and indistinct under the glare of natural light.

The cartoon—The cartoon is where construction begins. It is a plan containing all the information pertinent to the building of the window, such as the lead widths, decorative details, the panel size and the client's name. It stays with the window throughout assembly. The cartoon for the window you see on these pages is on 70-weight kraft paper (photo facing page), but any heavyweight paper that will take pencil lines can be used.

I begin a cartoon by marking it with the lines that define the full size of the panel. I get these dimensions by measuring the window opening from the inside of the rabbets, and subtracting ⅛ in. I note them on the cartoon next to the client's name, width first. This is when the window opening has to be checked for square, and any deviations noted.

After the full-size perimeter lines are drawn, I decide how wide the border lead should be. This is a function of how deep the rabbet is at the sash. I like about ¼ in. of lead line around the window, so I add that to the depth of the rabbet for the width of the border lead. For instance, if the rabbet is ⅜ in. deep, the border lead should be ⅝ in. wide. The dimension of the window from the inside edges of the border leads is called the daylight panel size.

Once I've marked the inside edge of the border lead on the cartoon, I draw in the positions of any other lead lines. These are all done at full size, and they are followed by more lines that mark the center, or heart, of the leads.

Patterns—Next, I sandwich a large piece of carbon paper between the cartoon and another piece of kraft paper. Tacks or tape hold both sheets securely to the table. I use a ballpoint pen to transfer the lead centerlines to the lower sheet. These become cutlines for glass patterns. During the copying, I occasionally lift a corner of the cartoon to see if I

From *Fine Homebuilding* magazine (October 1984) 23:36-41

missed any passages. Each shape represents a separate piece of glass, and each one is designated by location with a letter, as shown in the drawing at right. In a more complicated window, I designate the shapes with numbers. If there are colors in the window, they are also noted on the patterns.

I cut the lower sheet into a pile of patterns with a special pair of shears. Unlike a normal pair of scissors, this tool has two blades opposed by a third in the middle. The pattern shears remove a small fillet of paper as they cut, which compensates for the heart of the lead. There are different shears for different sizes of lead. The cut should be made with the portion of the blade closest to the rivet. The shears should be slid into the cut and only closed completely at the end of a line. Closing them in the middle of a cut will cause a nick that may snag the glass cutter later. The patterns have to be cut accurately, because they will reproduce exactly on the glass. Slight flat spots along a curve can be corrected, but in general the cut should split the carbon line. When the cutting is done, sort the patterns into piles by color and shape.

Cutting glass—Glass can be cut into simple or complex shapes by scoring one side and then applying concentrated pressure opposite the score marks. Glaziers use two types of tools to score glass for a cut: the diamond-tipped cutter and the steel wheel. The steel wheel is the more accessible tool and it's the one I used on this window. Steel wheels are inexpensive, and if they are cared for, they can handle the most complex of cuts. If the wheel is nicked, however, it will make a discontinuous score mark, which will make the break in the glass run wild.

The steel-wheel cutter should be stored in a container with some tissue at the bottom. This tissue should be soaked with a lubricant—mineral spirits, kerosene or any light machine oil. The lubricant is necessary for a successful cut. The oil not only helps the wheel spin freely during the cut, but also seems to affect how long the score mark remains open and breakable. A score is likely to heal over if the glass isn't broken within about 30 seconds.

When I make a cut against a straightedge, I first paint oil along the path of the cutter. This makes for a very clean break. Sometimes cranky old salvaged glass or the more brittle antiques and opals (see the sidebar at right) need this extra help.

The steel wheel feels quite awkward at first. Most beginners either push too hard or are indecisive. The score should be made on the smoother side of the glass. Hold the cutter plumb. The cut should be heard but not easily seen. If the cut is a fuzzy, sputtering line, too much pressure has been applied. Speed is not important but evenness is, so don't stop in mid-cut. You can let the wheel roll off the edge of the sheet to complete the cut without worrying about breaking the glass.

To make the break, hold the glass with a thumb on each side of the score at one end, and your index fingers bent and touching

The cartoon is a full-size drawing of the window with all the information pertinent to its construction. For this window, the border cames will be ⅝ in. wide, and the field cames ⅜ in. The lead lines are transferred, via carbon paper, to the bottom sheet. The shapes inside these lines represent the glass, and they are labeled, cut out and used as patterns.

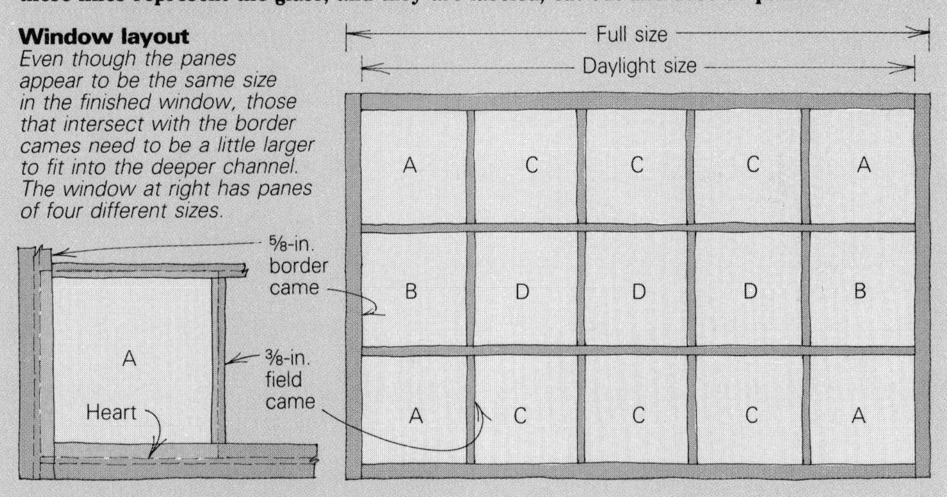

Art glass

Three basic kinds of colored glass are used most often in leaded windows. Antique glass is hand-blown in the form of a cigar-shaped bubble, which is then split and flattened in a kiln. It was dubbed antique by the 19th-century Gothic revivalists who are credited with developing the technique. Antique glass is watery, full of bubbles and surface imperfections, and it comes in more colors than any other kind of art glass. This is important in an art form where the pallet cannot be mixed. Blenko Glass in West Virginia makes this country's only antique glass. It is blown into wood molds, producing square bubbles with the imprint of charred logs on its surface. These small sheets, roughly 20 in. by 30 in., are usually easy to cut but they run very thick in places. They are available in a wide range of good colors from stained-glass supply houses.

Most antique glass is made in Europe. The English make small sheets of antique with many good medium and dark colors. They also cut well. French antique has a lot of internal tension and it can be brittle and unpredictable, but it comes in colors that are unavailable anywhere else, and the sheets are larger than most. German glass also comes in many good colors, and most of it is even-tempered. But some cuts erratically. As in most other types of glass, antique reds, pinks, yellows and oranges are more brittle than the other colors.

Flash glass is a type of antique with a clear or tinted base and a thin veneer of stronger color. It has to be cut on the side opposite the veneer. Most antique reds are flashed. Some machine-made glass goes by the title of semi-antique. It has some surface character but is generally flat and consistent in color. The best antiques will usually have uneven shading.

Opalescent glass was developed in America toward the end of the 19th century. The studios of Tiffany and La Farge used this variety widely, sprinkling their complex landscapes and plant motifs with the crucible-bred swirls, mottles, variegations and textures characteristic of this glass. Good opalescent glass is made in limited lots by small factories scattered around the U.S. At its best, it is lively and spontaneous. At its worst, it looks like colored plywood.

Cathedral glass, a machine-rolled commercial product, has a limited range of colors, a gravelly texture and a consistent thickness. It is inexpensive and boring. *—D. H.*

Illustrations: Frances Ashforth

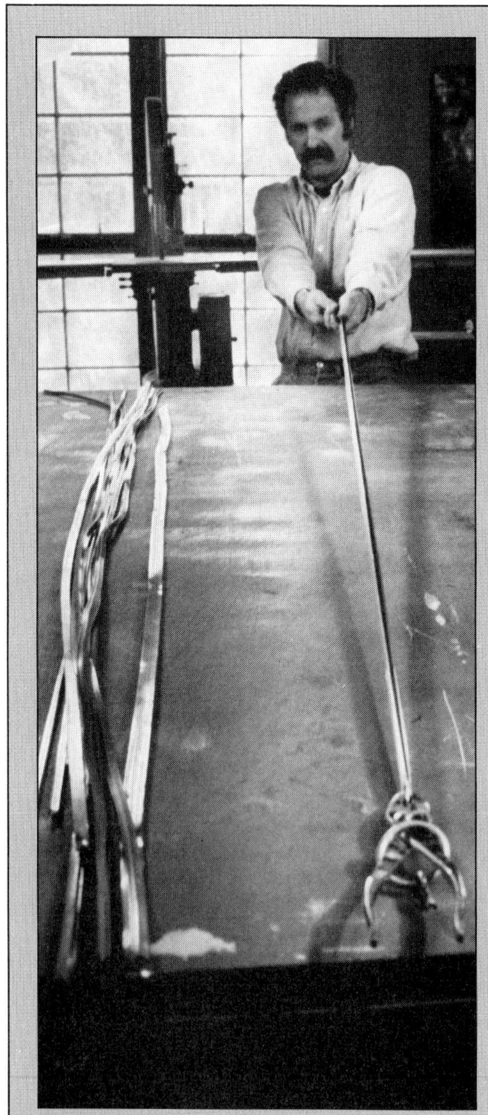

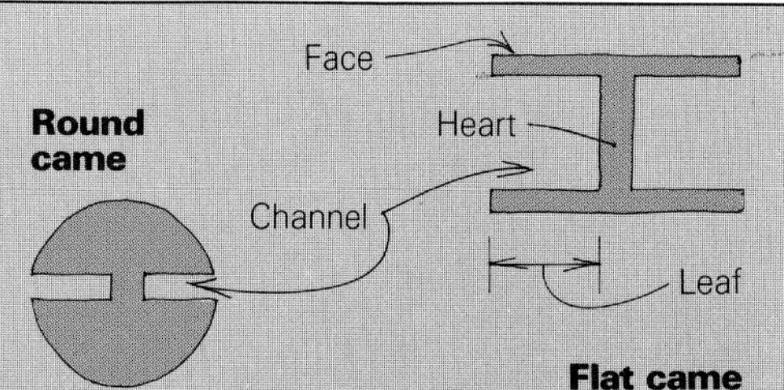

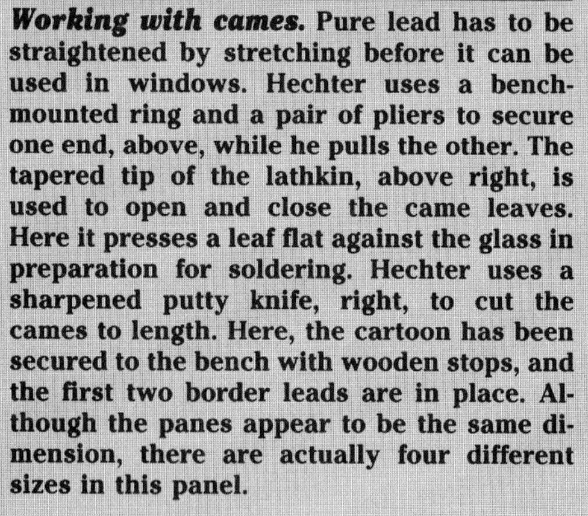
Working with cames. **Pure lead has to be straightened by stretching before it can be used in windows. Hechter uses a bench-mounted ring and a pair of pliers to secure one end, above, while he pulls the other. The tapered tip of the lathkin, above right, is used to open and close the came leaves. Here it presses a leaf flat against the glass in preparation for soldering. Hechter uses a sharpened putty knife, right, to cut the cames to length. Here, the cartoon has been secured to the bench with wooden stops, and the first two border leads are in place. Although the panes appear to be the same dimension, there are actually four different sizes in this panel.**

each other at both knuckles beneath the glass. As the thumbs press down, the index fingers become a fulcrum, and the glass will break along the weakened score line. On a stubborn piece or a tight curve, you can tap the score from beneath to start a crack. Use either the handle of the cutter or the grozing teeth found on the back side of the cutting end. This crack can then be chased along the whole cut, or used only to encourage a break. Breaking pliers can also be used to assist a stubborn cut. They have jaws up to 1 in. wide that meet only at the end of the tool. They are used to provide more leverage. But a tool-assisted break is never as clean as one done by hand.

Grozing is the term for nibbling away at a piece of glass. A notch on the cutter handle slips over the edge of the glass, and when the handle is levered upward, a bit of glass breaks off. This is how glass was shaped during the Middle Ages. Most grozing is done with a pair of grozing pliers. This blunt-nosed, untempered, parallel-jawed tool can be very useful at the cutting bench. It has jaws that meet only at the business end, and it is used as the name implies—when a cut can't be completed by hand. There is no inside curve that cannot be shaped by patient grozing, and I have won beer proving this.

To cut glass with the pattern, hold the pattern against the glass with one hand, and follow the edge of the paper with the cutter wheel. This method works for both curved and straight-sided shapes. If I'm going to make repetitive 90° cuts, such as the ones for the glass in this window, I set up a jig using a T-square and a stop, and I occasionally check the cuts against the pattern.

Working with cames—I'm often asked how I pour the lead between the panes. The lead is not, and never was, poured around the glass. It starts out cold as H-shaped strips called cames. They are about 6 ft. long, bend easily by hand, and can be cut with a knife. Cames may be less than ⅛ in. wide, or exceed 1 in. in width. The faces are usually round or flat (drawing, above left). Flat lead has a flexible leaf that closes down snug against the glass. The round style is usually heavier in section, with a stiff leaf not intended to be closed down. Round lead looks more delicate than flat lead, and it can be bent into tight, intricate curves. Most Victorian work uses round came.

Not all brands of came are pure lead. Antimony and other metals are being used more and more. These mixes are economical, and are not as prone to tarnish as pure lead. Many craftsmen prefer this new product because it solders so well, but its drawbacks make me hesitant to use it. It is harder and more brittle than pure lead, which makes it more difficult to cut and shortens its lifespan.

Lead cames must be stretched before they're used (photo far left). This hardens the lead, so I'm careful to pull just enough to straighten the came. Over-stretching also gives the surface a scaly texture. Alloy leads usually don't need stretching.

I used flat lead in this window, and the channels had to be opened with a lathkin. It looks like a fat knife blade (photo top center), and I use it constantly. Mine is 2 in. by 6 in. by ⅜ in. thick, and it's made of Teflon. Traditional lathkins are made of waxed hardwood.

Once the came is prepared for glazing, it should be set carefully aside. Its pliability causes it to kink readily, and if it's bent more than once through careless handling, it will lose its crisp look in the finished window.

In addition to the lathkin, I use a small hammer, a lead knife, a stopping knife and horseshoe nails. The lead knife I use is a hand-forged, high-carbon steel putty knife made in England under the brand name of Footprint. I have shortened the blade a bit and honed a sharp edge onto it. Many window makers use the German-style lead knife. It has a rounded cutting edge that ends in a hook. Its point is also useful for lifting the leads and poking around in tight places. I make a cut by setting the knife on the face of the lead, applying light pressure and rocking the knife back and forth (photo bottom left). After the blade penetrates the top leaves, the lead can be quickly chopped without crushing it. If many leads of a single length are needed, you can use a bandsaw or a radial-arm saw. Still, my everyday cutting is done by hand. The tools used in glazing have worked well for centuries, and there's little room for improvement.

Partner to my lead knife is my stopping

knife. Made from a wood-handled oyster knife, it is used for probing tight corners, lifting came leaves and prying. I cast a lead plug into its handle for tacking nails or tapping glass into place.

Glazing—This is the term used for the assembly of glass and lead into a window panel, and it takes place on a sturdy, flat workspace. The top of my glazing bench is a series of tightly fitting 2x planks, and it's 36 in. high. The cartoon is placed at one end, with ½-in. by 2-in. stops nailed along the bottom and the left-side lines (photo right). Check the stops with a framing square. If they're too long for the square, use the 3-4-5 method. After cutting and placing the border leads, I insert the first glass into the channels. The open edges of the glass should align precisely with the inside edge of the penline that transferred the design to the pattern. These lines will be the references throughout the glazing stage. If the glass is allowed to deviate from this line, the result will be distortions that will multiply as the panel progresses.

Flat came is tucked wherever it intersects. The flexible leaf is gently pried up to accept the intersecting lead. The hump that results is then carefully hammered flat. This tucking strengthens the window and enables the glazier to adjust the leadwork without recutting any of it. The tucked end should not extend all the way to the heart but should be held back slightly to prevent the panel from glazing full (expanding beyond the reference lines).

I tack horseshoe nails into the bench alongside each piece of glass as it goes into place. They pin the loose pieces against the stops until the cames are soldered, which happens when the entire panel is in place. As the window grows, I pull them up and stick them around the working perimeter.

Round-faced lead can not be tucked. It is always butted or mitered, and for it to be done right, the cuts have to be straight. If any undercutting occurs, what looks tight on one side will have gaps on the other.

Working from left to right, the quarry pattern is glazed one course at a time with one long lead capping each course. These leads can be kept straight by placing a stop alongside and tapping it with a hammer.

Occasionally, handmade glass runs too thick to fit into the channel. Many people avoid using it for precisely this reason. Seen from the outside, leaded glass has a certain faceted quality, and glass of random thickness contributes to this effect. Thicker glass also bends the light as it passes through the window, casting wavy patterns of sunlight around the room. Much new leaded glass is flat and inanimate because the craftsman holds flawless lead as the highest priority, using only glass that cuts and fits easily.

To modify the channel to accommodate thick glass, I sever the leaf from the heart with a pair of front-cutting nippers, the type that carpenters use for clipping nails. The leaf should be cut only on the side where the glass is uneven, leaving the other half intact. If necessary, I make several cuts ⅛ in. apart. This segmented leaf yields to the glass. It is then pressed flat against the glass and reconnected with solder. The soldering iron should be on the cool side. A scar remains, but the overall product is improved.

Glazing. **A window builds from left to right, with a long lead came capping each course. The short leads are tucked under the leaves of the intersecting cames, strengthening the panel. At the top of each course, the leads are adjusted with a few taps from a hammer against a stop. Note how the leads and the panes align with the cartoon.**

One further note on glazing: a neat, uncluttered bench shows up in the finished product. It is better to have unbroken leads wherever possible, for this reason. Leads partially woven into the unfinished panel will trail out onto the bench. They scratch easily, so don't let them mingle with tools, and can your lead scraps frequently.

Placing the last two border leads along the upper and right-hand edges completes the panel. Nail two more wooden stops to the bench, recheck your full-size dimensions and adjust the border leads accordingly. With the flat came there is a bit of give and take, and I slide the tucked leads in and out to adjust the size of the panel.

Once all of the joints have been hammered flat, the leads can be pressed down against the glass with the lathkin. The tool rides over the heart of the cames, and you have to be careful not to stress the glass. Individual leads can be straightened or aligned by placing the tip of the lead knife on the came at about a 45° angle, and then tapping its handle with a hammer.

Leads down and aligned, I affix a small pa-

Soldering. A special tip solders the cames, but before it can be used it must be heated, filed and tinned. Hechter's tinning tray is an old coffee can, above, and it contains a small amount of solder and powdered rosin or flux. The hot soldering tip is held slightly above the joint and a piece of wire solder placed between the two. The three are brought together for a few seconds, and the heat transferred through the solder to the came. When the solder melts, the tip is lifted and the finished joint looks like the one at right.

Finishing. Pigmented cement is forced into the cracks on the weather side of the window using a stiff bristle brush, left. In addition to sealing the window, the pigment tones down the brightness of the fresh lead. Excess cement is removed with brushes, rags and sawdust. Steel braces keep a leaded window from sagging or giving in to the wind. They are soldered to the interior side of the window. The ends of the braces are pounded flat into 'paddles' and soldered to the came with a chisel-tipped iron, above.

per label to the glass in the upper right-hand corner. This identifies the panel, and marks the interior side.

Soldering—A glazed panel should be soldered as soon as possible. If you wait, you get oxidation, which hampers the soldering. Minor oxidation can be cleaned away with steel wool or a wire brush. When the lead is badly oxidized, as is often the case with pure lead came exposed to moisture, brush the panel with flux and allow it to sit awhile before you scrub it. Be sure the panel is free of dust and debris before soldering. Any bits of lead or loose solder caught under a leaf will break the glass during cementing.

Oleic acid, a derivative of sheep fat, is the best flux for lead. With a brush, I coat each joint with flux just before soldering. Use the flux sparingly, because any excess will seep into the panel, form a puddle and then ooze out after the window is finished.

A soldering gun is of little use for all but the smallest of lead. The tool worth owning is a 250-watt American Beauty soldering iron with a ⅝-in. tip. Hexicon also makes a good iron.

The iron doesn't need a thermostat. Its temperature can be controlled by switching the power on and off. My iron is connected to a switched socket, and I know it's on when I see the red light out of the corner of my eye. The temperature demands for soldering leaded glass differ greatly from moment to moment, and the switch gives me the most flexibility.

For soldering the leads, I use a homemade tip made from a ⅝-in. piece of copper rod. It's bent at a 90° angle, and it has a blunt end. As soon as the iron starts to heat up, I file the tip smooth of the burrs and pits caused by the acid in the flux. I then tin the clean tip (coat it with a thin layer of solder) on the lid of an ordinary can (photo facing page, top left) on which is placed a little flux or powdered rosin and some solder. The tinning tray will draw excess heat off the iron, which makes it another means of temperature control. Before I start in on the panel, I test the iron on a piece of scrap lead.

Soldering done well requires a light touch, proper temperature and pacing. The tip hovers above the joint as a piece of 50/50 wire solder is placed between the two (photo facing page, top right). The solder transmits the heat as the tip is rolled slightly. When the lead has taken on sufficient heat, it will accept the solder and the joint will flatten. If the iron is too hot, the lead (which melts at a slightly higher temperature than the solder) will mingle with the puddled solder and make a lumpy joint. These can be flattened, but messing with them doesn't help them to look their best. When every joint has been soldered, I go back and check the panel for the joint that has inevitably been passed over. Then I wipe off the flux with a rag, pull the stops and dress down the unused channels on the outside of the border lead with the lathkin. The panel is now ready to turn over.

A half-soldered panel is still flimsy, and must be turned carefully. I pull it halfway off the bench and pivot it to vertical using the bench edge as a support. Then I lay it back down using the same technique in reverse. After truing any misaligned leads, I hammer the joints flat. This is the exterior side of the window, and the leads will not be closed down until the panel is cemented (explained below). Once the soldering is finished, I let the panel stand for a few days. This allows the flux to dry out. It should be tilted slightly, backed by boards for support. Larger panels should be stacked with a board in front as well to prevent them from folding over on themselves under their own weight.

Cementing—At this stage, the panel still rattles when it is moved, and it would leak on a dewy morning. It becomes a solid, impermeable unit when a waterproofing compound called cement is forced into the spaces between lead and glass on the outside of the panel. Cement is pigmented, and it turns the new metal shine of the panel to a dull grey. Quick cement can be had by mixing white gas and lampblack into a portion of steel-sash glazing putty. Stirred with an electric drill and paint-mixing attachment, it will flow freely, yet support a popsicle stick upright when its consistency is right. The best cement takes longer to concoct, but it is tenacious. The formula, in volume measurement, is: 12 units powdered whiting (powdered calcium carbonate), ½ unit powdered lampblack, 1½ units Japan or cobalt drier, ½ unit grey floor and deck enamel (alkyd resin type), 1¼ units boiled linseed oil, ½ unit turpentine. This compound is best when it's fresh, but it can be stirred up and used for a week if it's kept covered.

Gloves and an apron are a good idea for messy jobs like mixing and applying cement. The tools I use for cementing include one large and one small natural-bristle scrub brush, two good rags, sawdust, a stiff bent putty knife and a sharpened bamboo stick.

I ladle some cement to the open-lead side of the window, and work it into the cracks with the small brush (photo facing page, bottom left). Only one side is cemented because air pockets deep in the window will heat up under sun and blow out the seal of one side or the other if both are cemented. The one-sided approach works only with flat lead.

I brush the cement under the leads in a circular motion. Every space, every corner must be filled. A heavy stroke discharges cement from the brush, a light stroke pulls it off the window. After the entire side is covered, I lift excess cement off the panel with light strokes and scrape the brush on the side of the can. Then I lay the leads down with the bent putty knife, trapping the cement under the leaf. Ride the heart of the lead to avoid cracking the glass.

When the leads are down, I rub a handful of sawdust over the surface with a clean rag. Ideally, the rag itself should not touch the panel—just the sawdust, if you can manage it. This cleans the panel nicely.

Though the interior side of the panel does not get cemented, it needs a little color to break the shine of the lead. I rub the cement brush over this side, then clean up with the sawdust. To further clean the panel, I scrub both sides with the clean brush, working in the direction of the lead. Then I polish the panel with sawdust, applied with the clean rag, and I finish by removing excess cement along the cames with the bamboo stick. A window with round-faced lead should be cleaned with whiting instead of sawdust, which would embed itself in the open channel.

The cement takes at least a week to cure, so I let the panels rest before installation. I lean them up, and brace them so they don't bow. Straightening a bowed panel after the cement has set up will break the seal.

Bracing—Horizontal braces should be used every 18 in. to 24 in. on panels that are 12 in. wide or wider. Bracing stock is usually ⅜-in. or ½-in. by ⅛-in. galvanized flat bar. It is soldered on edge along a lead line on the inside surface of the window. Both ends of the bar should fall on a solder joint along the edge of the panel. Braces are cut with a hacksaw or bench shear to a length ⅜ in. short of the full-size dimension. The ends should be gently tapered, then flared out into hammered paddles (photo facing page, bottom right). The paddles enable the bar to make contact with the sash; this braces the window without having to modify the rabbet. The bar can be bent to conform to a curved lead, but the greater the bend, the weaker the brace will be.

Sometimes passing a bar over a section of glass can't be avoided. But the bars, which can look obtrusive during construction, usually become a lot less obvious when the window is finally in place.

To secure a bar, I first center it over the lead line I want it to follow. Then I paint tinner's fluid (flux for galvanized metal) on each side of the bar over every solder joint the bar intersects. The soldering iron must be fitted with a flat chisel-tip, and it must be very hot. The tip is loaded with solder and held to one side of the bar without touching the lead. I hold a solder wire to the other side of the bar, which will conduct enough heat to melt it. At this point the iron is punched down onto the lead, pulling the solder down and around the bar with it. The iron is hot enough to fry the lead and should not be in contact with it for long. The bar is held still until the solder is completely cooled. Next, I tin the open cuts to keep them from corroding, and then solder the paddles and bring them flush with the border lead. I remove any residual flux, first with a rag and then with newspaper and window cleaner. If a lot of flux is left, I spray glass cleaner on the affected areas and scrub them with a brush and sawdust. A little gun bluing (sold at most sporting-goods stores) brushed on the brace bar will eliminate its shine.

The window is now ready to install. If you built it right and install it properly, it should last for 500 years. □

Doug Hechter is a licensed glass contractor working in Santa Barbara, Calif.

Installing Glass Block

Privacy and light in the same package

by Michael Byrne

As a tilesetter, many of the jobs I've worked on over the years have included walls of glass block that were installed by others. Unlike bricks, masonry block and stone, which have rough textures and seem to suck up mortar and stay put, glass block slips and slides during installation, and refuses to behave unless treated correctly. Fed up with hours of making my tile fit around the lousy glass-block work I found, I began to learn that trade myself about eight years ago.

My first attempts to build a glass-block panel (a panel is a "window" of block surrounded by wall) seemed awkward and confounding, and before long, I began to sympathize with those whose work I had criticized. But I've done a number of glass-block jobs since and am finally confident in my skills. What follows are some tips that will help you to avoid the painful learning process I went through.

Characteristics of glass blocks—Glass blocks were originally made by hand (sidebar, facing page), but today's blocks are machine-made by pressing two molded halves of semi-molten glass together. The edges of the inner and outer faces create a flange that allows the block to key into the mortar joints, and allows space for metal reinforcing rods or wire. The raised bump formed at the point where the two halves are fused together is an additional mortar key. To reduce the slipperiness of glass when mortar is buttered on, the hidden edges of blocks are spray-coated with a plastic-like polyvinyl butyral coating while they are still hot from fusing.

The most common sizes of hollow glass block are 6x6, 8x8 and 12x12. Other sizes and shapes are available, including 4x8 and 6x8 rectangular blocks, and hexagonal corner blocks. Blocks manufactured in the U. S. are modular; that is, the nominal size includes an allowance for a ¼-in. mortar joint. The actual dimension of a modular block is ¼ in. less than the nominal size (a 6-in. block, for example, is actually 5¾ in. square). Though you can find a few glass artisans who will provide custom-made solid glass block, there's only one company in the U. S. that manufactures glass block—Pittsburgh-Corning Corporation (800 Presque Isle Dr., Pittsburgh, Pa. 15239). Imported glass blocks are available (see the list on p. 97), but direct technical information and assistance may be limited.

In addition to the different sizes, Pittsburgh-Corning makes hollow block in two thicknesses. The standard block is 3⅞ in. thick and has an insulating value of R-1.96. It can be used for both interior and exterior applications. An 8x8 block weighs 6 lb. For light commercial and residential jobs, Pittsburgh-Corning makes a "Thinline" block that's 3⅛ in. thick, with an insulating value of R-1.75. Because Thinlines aren't as strong as the thicker blocks, their use is limited to exterior panels of 85 sq. ft. and interior panels of 150 sq. ft. But they are about 20% lighter than standard blocks, so they're ideal for situations where weight would be a problem. An 8x8 Thinline weighs 5 lb. Another kind of glass block is solid, instead of hollow. Called VISTABRIK, it's a solid chunk of glass that's suitable for translucent pavers between two levels of a structure, or as a nearly vandalproof panel. An 8x8 VISTABRIK weighs 15 lb., and can withstand a 30.06 rifle shot from 25 ft.

Standard glass blocks have a compressive strength of 400 psi to 600 psi, and solid glass VISTABRIK blocks are rated at 80,000 psi. Glass blocks, however, should not be used in load-bearing situations, regardless of the application or panel size. Provision must be made to channel loads around the block panels with properly sized lintels or headers.

Most of Pittsburgh-Corning's line can be specially ordered with inserts of translucent fiberglass that improve the R-value by about 5%. You can also get solar reflective block that cuts down on solar heat gain. In combination with these characteristics, various patterns can be pressed into the inside surfaces during manufacture, while the two halves are still semi-molten. These patterns change the way light passes through the block, and also change the amount of privacy blocks offer. Some blocks are clear, while others admit light but not the view. Putting the pattern on the inside of the block leaves the outside of the block smooth, making the blocks easier to clean once they're in place.

Mortar and admixes—Glass blocks are stacked one upon the other, and mortar joints are usually aligned both horizontally and vertically. You can stagger the joints if you wish, but this makes the blocks more expensive to install. Laying up glass block is like laying up concrete block or brick, except for one crucial difference: the consistency of the mortar.

Masons work with a rather moist mortar mix to compensate for the absorbent nature of brick and concrete block. When either one is laid up, moisture is wicked from the fresh mortar, which then stiffens and supports the weight of the masonry. Glass blocks, however, aren't absorbent, so standard mortar won't stiffen as quickly. Instead, it will ooze out of bed joints and head joints after a few blocks are in place—a sight greeted with considerable dismay by novices.

Mix your mortar on the dry side. Pittsburgh-Corning recommends a mix of 1 part portland cement, 4 parts clean sharp sand and ½ part lime (measure by volume). Add just enough water to change the mix from crumbly to spreadable—roughly five gallons of water to a 94-lb. sack of cement. The mix should just barely wet the edges of the blocks. (I prefer a mortar mix of 3 parts sand and 1 part cement, no lime, with a liquid additive in place of water. I'll discuss additives more below.)

You won't have a lot of working time with the drier mortar, so in moderate temperatures mix only what you'll need in about a half-hour; mix even smaller batches in very dry or hot weather. Don't add more water to mortar that begins to set up because this prevents it from curing properly. It also increases the likelihood of mortar cracking when it sets up.

You can use a pre-packaged mortar mix to which you add only water, though I prefer to mix my own. These mixes usually set up faster than a standard mortar, which can sometimes be an advantage. Pre-packaged mortar mixes have a disadvantage, too. They often contain a finer grade of sand than you get from a building-supply yard, so the mortar will shrink more as it sets. On small panels this won't present a problem, but on larger jobs you should use some sort of liquid mortar additive to control shrinkage.

The additive should also provide some waterproofing. As one side effect of waterproofing, the mortar gets stickier and so gets a better grip on the slippery blocks. An additive also increases the flexibility of the cured mortar joint, reducing cracking. The additive I use (which does all this) is Laticrete 8510 (Laticrete International Inc., #1 Laticrete Park North, Bethany, Conn. 06525). If you're making mortar from scratch in an area where special additives are hard to come by, you can get a sticky mix by using waterproof portland cement instead of standard portland.

A third kind of additive, called a mortar fortifier or lattice, can increase the compressive strength of the mortar. It will also make the mortar stickier and more flexible, and increase its water resistance after cure.

Panel design—The overall size of a panel and whether it's to be built on the interior or exterior of your house will determine the thickness of

From *Fine Homebuilding* magazine (February 1987) 37:46-51

In this shower surround, glass block was used as an interior partition and as a divider between the shower and the adjacent sunroom. The pattern inside each block obscures views yet admits light.

block to use and the method of installation. Though the design of glass-block walls can get complicated on large commercial installations, most residential jobs are pretty straightforward. For design purposes, these fall into two categories: small panels (under about 25 sq. ft. in area) and large panels (up to 144 sq. ft. in area).

Small panels, like a slender sidelite alongside an entry door or a glass-block divider in a shower room, add a touch of elegance and style to a house. From a technical standpoint, a small interior panel, like a block window between two rooms, is the easiest to build. Jobs this small can be mortared in tight at jambs and headers. For interior panels, no reinforcing is necessary, and you needn't waterproof the surrounding framework. But make sure that even a small job is adequately supported. Glass blocks may look light and airy, but a small panel of standard 12x12 blocks can weigh as much as 400 lb., not including the mortar. Put in extra cripples beneath the sill, or build up a heavyweight sill with an extra 2x to keep deflection to a minimum.

For a small exterior panel (and any other panel frequently exposed to water), the sills should be waterproofed before the block is installed. Pittsburgh-Corning doesn't call for waterproofing the jambs, but I often do it anyway. Tar paper embedded in asphalt emulsion is a suitable waterproofer for many situations. But I prefer to use a waterproof membrane (like Nobleseal, made by The Noble Co., 614 Monroe St., Grand Haven, Mich. 49417). Membranes are fairly easy to install, too; just wrap a sheet 2 in. or 3 in. over the framing and staple the edges in place; lap the seams and seal them with Noble adhesive. Even though Pittsburgh-Corning requires no reinforcing for a small exterior panel, I feel more comfortable about the job if it is reinforced. I'll talk more about reinforcing below.

For large interior panels, like a full-height partition between two rooms, the design gets a bit

Substance and shadow

The history of plate glass and glass block is linked with the history of modern architecture. Early in this century, the classic precepts that underpinned much of architectural thinking were being challenged, and a fresh new vision was emerging in all the arts. This vision was due, in part, to the technical advances in construction materials and techniques that allowed architects more flexibility with structural design. Cast-iron columns could be spaced widely apart, and because of advances in glass technology, ever wider and longer sheets of plate glass began to fill the spaces between.

Glass block was another product whose invention depended largely on manufacturing advances. The "glass brick" walls in early modern architecture were actually single thicknesses of pressed glass squares set in large reinforced-concrete panels. These panels were either formed on site or precast before being installed as infill between structural elements. Auguste Perret's Notre Dame de le Raincy (1922) and Le Corbusier's Immeuble Porte Molitor (1933) provide examples of this technique.

In 1902, the Corning-Steuben Company invented the hollow, modular glass units that became known as glass block. The blocks were made by pressing molten glass into identical molds. The square "dishes" thus formed were sealed together at high temperatures, with a partial vacuum forming in the space between dishes as hot air cooled and contracted. Earlier attempts to form glass blocks using traditional glass-blowing techniques had proven unsatisfactory, because moist air from the lungs of the glass-blowers would condense on the inside of the blocks, clouding them. Owens-Illinois made glass block at one time, but with architectural preference for materials nearly as changeable as Paris fashions, they dropped out of the market years ago.

The directions for installing glass block offered to builders in 1902 were amazingly similar to the specifications offered today by Pittsburgh-Corning (formed by the merger of the Corning Glass Works and Pittsburgh Plate Glass Company), which is the sole U. S. producer of glass block. Back in 1902, Corning-Steuben said "The [glass] bricks are laid in a similar way to ordinary bricks, with a mortar consisting of one part portland cement, two or three parts fine sharp sand and one-fifth lime mixed not too thin with water." Contemporary Pittsburgh-Corning instructions go on to suggest the addition of a waterproofing admixture if waterproof portland cement mortar is not used.

The revival of interest in daylighting is one reason for the current wave of interest in glass block. Another factor, ironically, is a revived taste for historic allusion in architecture. European and Japanese manufacturers are now scrambling to catch a share of expanding sales in glass block.

Arthur Korn, in his 1926 book *Glass in Modern Architecture,* spoke for all the pioneers of modern architecture when he identified glass and glass block as altogether exceptional materials, "at once reality and illusion, substance and shadow."

—Ronald W. Haase

Any glass-block panel exposed to water, such as this divider between a tub and a shower stall, should be treated like an exterior job. Jambs and sills should be waterproofed. In this case, tile fills that role. Top left: A block, with one edge already buttered with mortar, is being set onto the mortar bed. The wood strip on the wall acts as a guide to keep the blocks plumb. Top right: Additional rows of glass block are buttered on one edge and pressed into a layer of fresh mortar. In this case white thinset was smeared on the wall to increase the bond between the mortar and the wall. Once the blocks are in place and the joints have been tooled, any mortar haze should be removed with a sponge and plenty of clean water, as shown at right. Ceramic tile caps the edges of the blocks. The drawing below shows typical installation details.

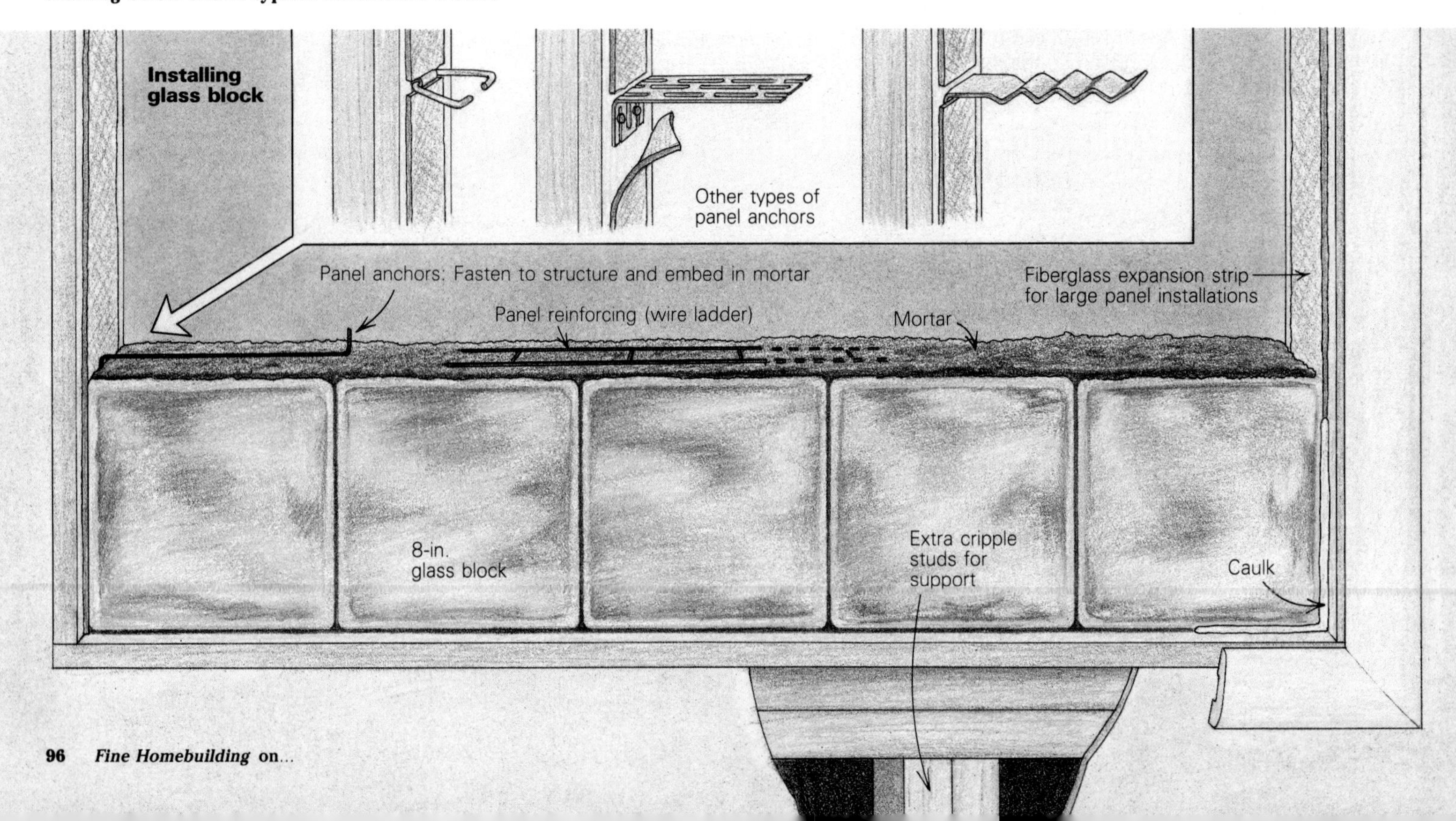

trickier. The panel must be isolated from the surrounding framing to allow for expansion, contraction or deflection of the framing members. But at the same time, it must be anchored securely to the building. To satisfy both conditions, various kinds of expansion strips and special metal panel anchors can be used to anchor panels to a building (drawing, facing page). One is nailed into the wall and the other is embedded in the mortar.

The use of expansion strips at jambs and headers is essential because they allow the surrounding framing to expand and contract without destroying the panel. The block itself won't move much unless the panel is very large; glass block's thermal expansion rate is .0000047 in. per degree Fahrenheit. Expansion strips are usually 4⅛ in. wide, ⅜ in. thick and 2 ft. long, and are made of dense fiberglass or polyethylene. They can be stapled, nailed or glued in place.

Where the edges of the expansion strips are exposed, fill the gap between wall and panel with packing. A common type of packing, called backer rod, looks something like a rope of stiff foam, and is just pushed into the joint. I prefer to use packing material that's square or rectangular in cross section, like POLY-VOID (Stegmeier Corp., 750 Garcia Ave., Pittsburg, Calif. 94565). Remember to take the thickness of expansion strips into consideration when sizing openings. A bead of caulk covers the packing.

Large exterior panels should be installed like large interior panels, but with the additional precaution of waterproofing. Because there are a number of variables to consider in the design of any large panel, I'd suggest that you consult with your local distributor, who may in turn contact a technical representative at Pittsburgh-Corning.

Reinforcing—Like masonry walls, walls of glass block require a certain amount of metal reinforcing to resist bending stresses. Pittsburgh-Corning offers panel reinforcing "ladders" that consist of two parallel runs of stiff wire that are separated by cross wires. The reinforcing should be placed in continuous rows every third course in standard block walls, and every other course in Thinline and VISTABRIK walls.

Proper reinforcing in the mortar joint is essential on larger jobs and useful on smaller ones, but not every glass-block outlet carries the full line of accessories. When scheduling problems won't allow time for tracking them down, I rely on an old standby to fortify small installations. In a pinch, 9-ga. galvanized wire can serve as both reinforcing and anchoring on small panels. For large panels, I'd definitely stick with the standard Pittsburgh-Corning reinforcing.

Installing a panel—Not long ago, I did a project that involved several different panels of glass block. The block I used for all three panels was 12x12 VUE, by Pittsburgh-Corning. I'll show you how I did the smallest one, a divider between the shower and the tub, because it illustrates the versatility of glass block (photos facing page). As you can see, a block panel doesn't always have to be entirely enclosed by wall.

Earlier I had installed ceramic tile on a wood framework that wrapped around the shower, along with two vertical lengths of rebar to stabilize the "open" side of the panel. Later the rebar would be covered with more ceramic tile. Because the sill was already waterproof, I dispensed with what is normally the first step in such a job: giving the sill a thick coating of asphalt emulsion (normally, this should be allowed to dry before any mortar is applied) or flashing it with a suitable waterproof membrane.

I glued expansion strips to the wall jambs with asphalt emulsion, and set a wood guide strip along the wall to keep the blocks running true. Such a guide speeds the work by giving me a constant reference to plumb. Panels should be anchored to the wall just above the first course, just below the last course, and at some intermediate courses (every 24 in. for standard block, and every 16 in. for Thinline). For this panel, L-shaped lengths of 9-ga. wire, fastened to the wall with 10d nails, were sufficient to serve both as panel reinforcing and anchoring.

When I was ready to lay the block, I mixed up a batch of mortar on a mudboard, scooped up a trowelful of mortar and slid it onto the sill, using a sweeping motion to spread it. All layers of mortar (the mortar bed) should be a full thickness, not furrowed. The slick edges of glass block need 100% support on this first course, so they must be bedded completely in the mortar. The first block went against the expansion strip and, as with all the remaining blocks, I gave it a couple of raps on the top with my fist or a rubber mallet to seat it securely. The block should butt up tight against the expansion strip, but should not be mortared here. The next block was buttered with mortar on one edge, then slid against the first and tapped in place. If you were doing a longer row of blocks you'd just continue this routine of butter, tap, butter until you reached the end of the first row. After every few blocks I carefully check for plumb and level, using a small level, and cut away any excess mortar from the blocks with the edge of a trowel.

Reinforcing should be centered in the mortar bed and run the full width of the panel. When I have to use more than one length of reinforcing to reach the end of the row, I overlap the pieces at least 6 in. or so. The panel anchors should be placed directly over the reinforcing, but they don't need to be fastened together; mortar provides the connection between them. I bend the anchors so that 6 in. to 8 in. contacts the jamb.

While stacking the rows of block, be sure to maintain plumb and level. Strings tacked to matching layouts above and below the panel are especially helpful when laying up curved or serpentine panels. The important thing to remember about glass-block walls is that light will be passing through the finished panels to highlight any inconsistencies in the mortar joints. Very slight variations in joint thickness are acceptable, but goofs or sloppy work practically scream for attention.

It's sometimes hard to maintain a consistent mortar-joint thickness, particularly if the installation is a tricky one, or if you're a novice. One trick is to use wood spacers cut from scrap stock. Wiggle them into the mortar bed just before setting a course of block, and they'll support the blocks while the mortar sets up. Soak the spacers before you put them in so they don't suck moisture from the mortar, and stuff the holes that remain with mortar.

Finishing the job—When the mortar has stiffened a bit, the joints should be tooled. On a large project you can't wait until all the block is up before tooling the joints, but usually you can on a small panel like this one. When the last row is in place, I step back and look at the joints. I usually find voids that need a little extra mortar, so I fill them in. When the voids are filled, each joint can be smoothed with a striking tool, a metal bar with a C-shaped cross section (available from masonry-supply outlets). This forces the mortar into any voids between the blocks, and also it compacts it to make it harder and more waterproof. If you can't get a striking tool, you can use a length of smooth copper pipe.

After tooling the joints on both sides of the panel, I clean up each block with a wet brush or a sponge, rinsing it frequently. If you have ever grouted tile, you will undoubtedly find this step familiar. Any lingering light haze can be removed with a piece of cheesecloth.

On some jobs, to add a little color to the transparent wall, I'll rake out the joints to a depth of ¼ in. to ⅜ in. and fill the resulting channel with colored grout, which I then finish smooth with a striking tool and clean as above.

Glass blocks can be tricky to install properly, but they can also be loads of fun. Just wait for a winter's day when the snow is piled up around the house and the sun is low in the sky. That's the time to pull out the old lawn chair, stretch out and close your eyes—you're as good as on the beach in Jamaica. □

Michael Byrne is a tilesetter in South Hero, Vt. Photos by the author, except where noted. His book, Setting Ceramic Tile, *will by published by The Taunton Press in the fall of 1987.*

Sources of glass block

Only one company in the U. S. manufactures glass block on a commercial scale:

Pittsburgh-Corning Corporation
800 Presque Isle Dr.
Pittsburgh, Pa. 15239.

Other sources of glass-block:

Forms & Surfaces
Box 5215
Santa Barbara, Calif. 93108
(Japanese glass-block)

Euroglass Corp.
123 Main St., Suite 920
White Plains, N. Y. 10601
(French glass-block).

Two companies import glass block from West German manufacturers:

Solaris U. S. A.
Division of Sholton Associates
6915 S. W. 57th Ave.
Coral Gables, Fla. 33143

Glasshaus, Inc.
P.O. Box 517
Elk Grove Village, Ill. 60007.

Greenhouse Shutters

Insulated panels and rigging that folds them away overhead

by Stan Griskivich

When my wife and I built our house in 1978, to cut heat loss I made and installed thermal shutters that fold back away from the windows. When we finally added a greenhouse last year, I knew we'd need insulation for its sloping south-facing glass, too. But shutters hinged on sloping studs would be awkward to operate, and would get in the way of people and plants when folded back. Insulated quilts don't work well on sloping windows larger than about 2 ft. by 6 ft. either, because their own weight pulls them out of their tracks or away from their magnetic strips. So I developed an overhead bifold arrangement operated by a line-and-pulley system adapted from nautical rigging. It takes just a few minutes each day to raise and lower the shutters.

Stan Griskivich is a carpenter and cabinetmaker who lives near Yarmouth, Maine.

The shutters, which fit into the slope of the wood-frame wall, incorporate a Thermax core and two dead-air spaces. They give double-glazed windows an R-value of about 14, as opposed to the R-8 of the house shutters, which have no Thermax core, and only one dead-air space. Because so much of a greenhouse is glass, it is important for greenhouse shutters to have a high R-value if it can be done cost-effectively. With my shutters it can be. Owner-built with the tools and materials listed on the facing page, they cost about $3 per sq. ft.

Measuring and cutting—Measure the width and height of your windows. Subtract ½ in. to determine the size of your shutter sets. These fold horizontally, so divide the set's height in half to get the length of each shutter's stiles (vertical frame members). I use a rabbet joint at the corners, so I subtract 1½ in. from the shutter set's width to find the length of the rails (horizontal frame members).

Rip 5/4 by 6-in. pine boards into 1¾-in. wide rails, and then cut them to length. If your south wall slopes, as mine does, you'll have to cut the bottom end of the stiles and the two sides of the rail on the lower shutter at an angle to match that of the slope. I used a sliding bevel to find the angle, and ripped the rails on my table saw and crosscut the stiles on my radial arm saw. Plow a ¼-in. deep by ½-in. wide groove down the inside face of all stiles and rails. For the groove in the bottom rail of the lower shutter, set the table saw just as you did to cut the edges of the rail to the proper angle. The ½-in. Thermax core fits into this slot. Now cut the rabbet ¼ in. deep by 1 in. wide in the ends of the stiles.

Cut the Thermoply panels to the same size as the shutters, two panels per shutter. I use a fine-tooth veneer blade on a table saw. A utility knife and straightedge also work fine.

Cut the ½-in. Thermax board 1½-in. shorter and narrower than the shutter. Be sure that the Thermoply and the Thermax panels are cut squarely and accurately.

Assembly—Put yellow (aliphatic-resin) glue on the ends of one rail and fasten stiles to rails with 2-in. drywall screws. Slide the Thermax board into the grooves of the stiles, then install the other rail. You now have a wood frame the same size as a shutter with a Thermax core recessed ⅝ in. deep within it.

The upper shutters require a pulley in each rail. To support them, glue ⅝-in. thick blocks to the center of their window sides, as shown. Thermoply will cover the blocks.

Run a bead of panel adhesive around the

Shutters prevent radiant and conductive heat loss through glazing. Here the author lowers a pair of his folding greenhouse shutters.

From *Fine Homebuilding* magazine (February 1983) 13:48-49

perimeter of the completed frame and lay down a sheet of Thermoply, foil face in. Fasten it with ½-in. to ⅝-in. staples set flush with the surface, about 8 in. o.c. Turn the shutter over and install the other sheet of Thermoply the same way. If necessary, sand the edge of the Thermoply flush with the surface of the frame. Then paint or seal the edges of the frame to prevent moisture absorption. The white surface of the Thermoply can be left as is, painted, stenciled, or covered with fabric or wallpaper.

Hinges and rigging—When you have completed both shutters for a single folding unit, join them with a pair of 3-in. by 3-in. loose-pin hinges. Locate the hinges with their barrels facing the window, and make sure that you leave 1-in. to 1⅛-in. clearance for the pulley, which would otherwise keep the shutters from folding completely. Once the hinges are in place, separate the two shutters by pulling out the hinge pins. Now install hinges on the top horizontal rail of the upper shutters—this time with their barrels away from the windows. The shutter pair will hang from the top plate of the south wall.

Measure across the ceiling a distance equal to half the height of the windows, and screw a 1x3 or 1x4 along the length of the greenhouse to support ceiling pulleys.

The rigging (drawing, top right) is easy to understand if you have ever been sailing. Begin by installing a pulley (I use Nicro Fico NF 414) on the ceiling strip, directly behind the vertical center of the shutter set. Now install another NF 414 pulley in the upper shutter's top-rail screw block, and an NF 219 pulley in the screw block on the bottom rail.

This part is easier with two people. Mark the location of the hinges on the top plate by holding the upper shutter against the stops with its pulleys facing the windows. Leave equal clearance on both sides of the shutter.

After the upper shutter is hinged to the top plate, re-connect the lower shutter. Drill a ¼-in. hole through the bottom rail of the lower shutter. Run a ³⁄₁₆-in. braided nylon line through the hole from the window side, and tie a stop-knot on the inside surface. Run the line to the NF 414 pulley in the upper shutter, then down to the NF 219 pulley at the meeting rail. Drill another ¼-in. hole through the shutter and run the line to the second NF 414 pulley, mounted on the ceiling. Once this rigging is complete, pulling on the line will simultaneously fold the shutters and draw the pair against the ceiling. A cleat fastened to the back wall secures them out of the way.

When you want to lower the shutters, let them down from the ceiling until you can reach up and begin to unfold them. From this point they will unfold by themselves as you pay out the line.

Install 3M V-Seal weatherstrip on the 1x4 glazing stops and the top, bottom and meeting rails. Install 1½-in. by 3½-in. turn buttons at the midpoint and bottom all of the sloping 2x6 studs; they will hold each set of neighboring shutters closed. □

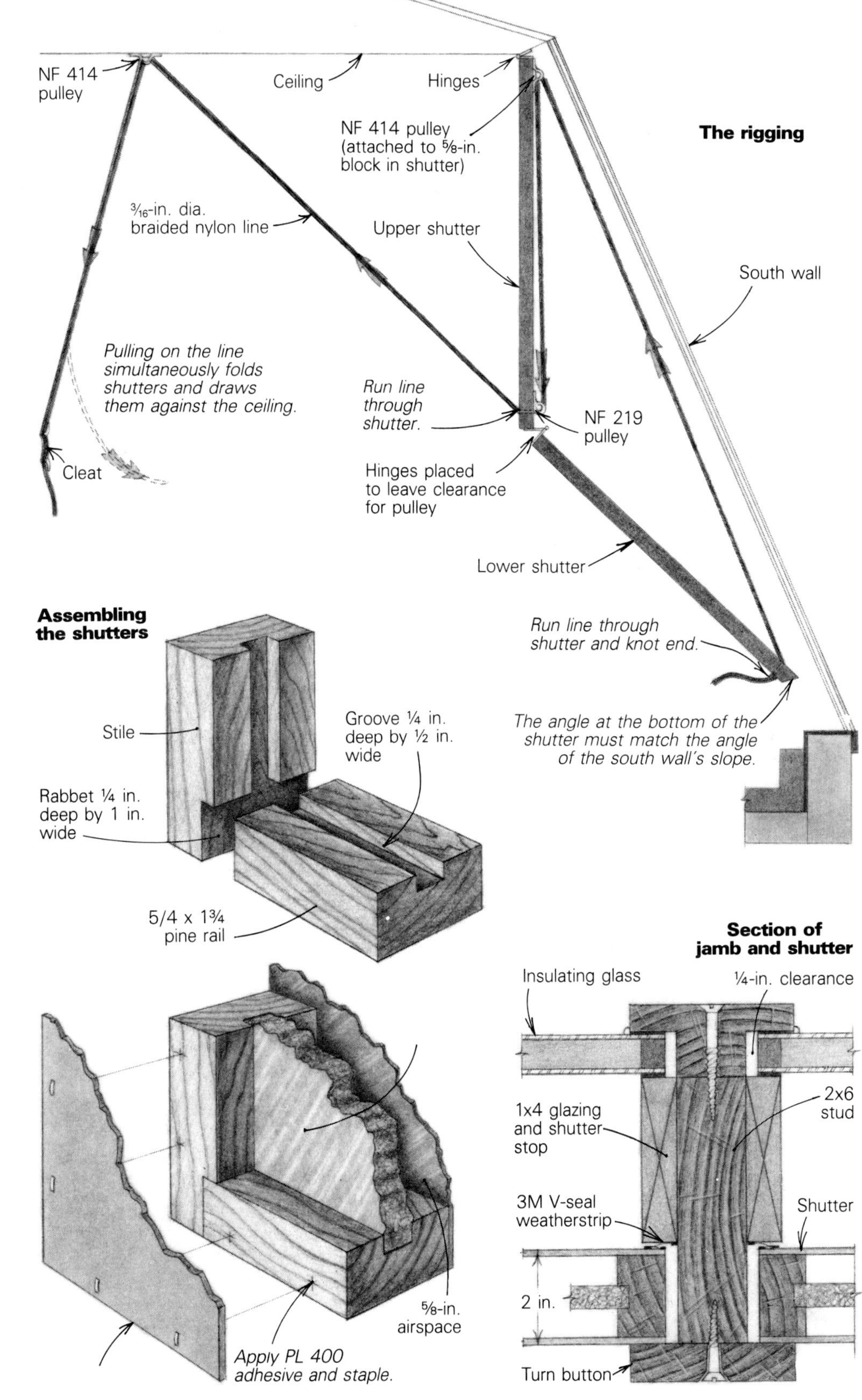

Materials

⅛-in. Farmline Thermoply (Simplex Products, Box 10, Adrian, Mich. 49221). Also known as White Face Thermoply, it comes in 4x8 sheets with one foil face and one white.
½-in. foil-face Thermax insulation board (Celotex Corp., Tampa, Fla. 33622), available in 4x8 and 4x9 sheets.
5/4 x 1¾-in. pine (ripped from 6-in. boards)
Contech PL 400 adhesive (caulking-gun cartridges)
3-in. by 3-in. loose-pin hinges
3M #2743 V-seal weatherstripping
Nicro Fico NF 414 and NF 219 pulleys (Nicro Corp, 2065 West Ave. 140th, San Leandro, Calif. 94577)
³⁄₁₆-in. dia. braided nylon line
4½-in. aluminum cleats
½-in. to ⅝-in. staples
2-in. drywall screws
Yellow (aliphatic-resin) wood glue

Tools

Table saw or radial arm saw with dado head
Utility knife and straightedge or fine-tooth plywood blade
Electric or hand stapler
Electric or hand drill and ¼-in. bit
Chisel or router (for hinges)
Screwdriver, awl, caulking gun, tape measure

Illustration: E. Marino III

Movable Insulation for Skylights

High insulation value and ease of operation ensure maximum energy returns from your skylights

by Larry Medinger

Skylights, along with properly sized south-facing glass, take best advantage of the sun in climates where winters are mostly cloudy—Oregon, in this case. But finding a convenient and effective way to keep sunspaces from overheating in the summer and losing heat in the winter is a problem. The solution shown here is to mount insulated panels in tracks so that they can fully or partially block the skylight openings. Moving the panels is just a matter of a slight tug on a rope.

Skylights are an asset to any home, particularly in cloudy winter climates. They can be the major source of solar heat gain during most weather, and they can make any room feel light and airy on the gloomiest of days.

Skylights will collect more direct sunlight in months when vertical windows will collect less, and the two glazing systems can reinforce one another to provide a balanced influx of sunlight all year long. Skylights will allow sunlight to penetrate deep into the structure, where it is less expensive and intrusive to the home's design to place thermal mass.

There are many areas in the United States where cloudy winter weather predominates over clear sunny days. In the typical textbook scenario, the low-angle winter sun shines brightly in under the 2-ft. eave overhang and charges the thermal mass. The reality is that over much of the country, for much of the winter, solar radiation comes essentially vertically from the cloud cover overhead.

South-facing vertical glass under a 2-ft. overhang will collect little of the available radiation on cloudy days. For the purposes of daylighting in such conditions, skylights work better than south-facing glass, and they can help collect more of the spare solar heat on these days.

The challenge is to create a comprehensive, climate-sensitive glazing scheme that will be properly sized with the heating and daylighting needs of the house. It must be correctly installed and flashed, and have an insulation system to protect against heat loss on winter nights and to keep the spaces below from overheating on summer days. To provide this kind of movable insulation for skylights, we developed a way to build easily operated, high R-factor, insulating panels for "sky-facing windows."

A first attempt—A number of the homes we've built in the Pacific Northwest use banks of skylights in one or more centrally located collection rooms. Each skylight is mounted between two beams on the ceiling. Our first design featured sliding panels made of rigid-foam insulation in wood frames. The sliding panels were supported between the beams by projecting wooden tracks. The panels could be raised and lowered along the ceiling by a counterweighted system of ropes and pulleys. The contraption worked, but the ropes and wood-to-wood contact of panel and track produced too much friction and made the panels difficult to raise. What we needed was an insulated panel that could be

From *Fine Homebuilding* magazine (August 1985) 28:46-48

moved easily into place and back out of the way, and one that we could make from off-the-shelf parts.

New tracks—For the tracking system, we finally settled on standard bypass sliding closet-door hardware. We used the wheels without modification. But bypass steel tracks (we used Cox #12-100, which are readily available in hardware stores) accommodate a pair of doors in the same opening. We needed to divide them into two single tracks. Using an abrasive cutoff wheel, we ripped the track into two pieces on our table saw. This operation demands wearing a face shield, as lots of shrapnel from the cutting wheel and the track gets hurled during the cut.

The object is to rip the track exactly in the right place (detail drawing, bottom of page). To keep the track from wandering away from the fence during the cut, you need to clamp a board to the saw table ahead of the blade to hold the track firmly and consistently against the fence. Next, grind or file the sharp edges and dings off the sawn edges to make them reasonably smooth. Sometimes in the ripping process, the tracks take on a slight curve. Be sure to straighten them before proceeding. Then drill ⅛-in. dia. holes in the tracks every 8 in. to 12 in. to accept Phillips-head 1¼-in. drywall screws. The tracks and other hardware are then cleaned with vinegar and spray-painted flat black.

Another problem we had with our first wooden track system was that it was difficult to achieve a durable seal between the ceiling and the panel in its closed position. A good seal is important because it reduces air movement, and thus temperature exchange, between the chilled skylight well and the warm room. It also inhibits condensation on the inner surface of the skylight. While the ends of the panels can be sealed by being made to contact weatherstripped bumpers, the sides have to slide past the surfaces they should be sealing tightly against. This causes friction and wear.

We found a workable solution by mounting the bypass tracks on the beams at a slant to the ceiling, as shown in the photo and drawing above right. The panels themselves are actually kept parallel to the ceiling by mounting the front and rear wheels at different attitudes. This allows the panel to move freely until the moment that it contacts the bottom edges of the skylight well. The pressure of the counterweight or the weight of the panel (depending on whether its closed position is uphill or downhill) keeps a positive contact between the well and the panel, and so maintains a good seal between the two. We have used either commonly available foam weatherstrip gaskets or light-colored fabric rolled under on the edges and stapled to the tops of the panels.

Making the panels—We build our panel frames of clear vertical-grain Douglas fir. For the panel itself, we usually use Masonite and glue it into grooves in the surrounding frame. We expose the smooth side to the room and paint it the color of the ceiling or walls. The frame is finished the same as the nearby beams and the rest of the trim in the room. The insulation in-

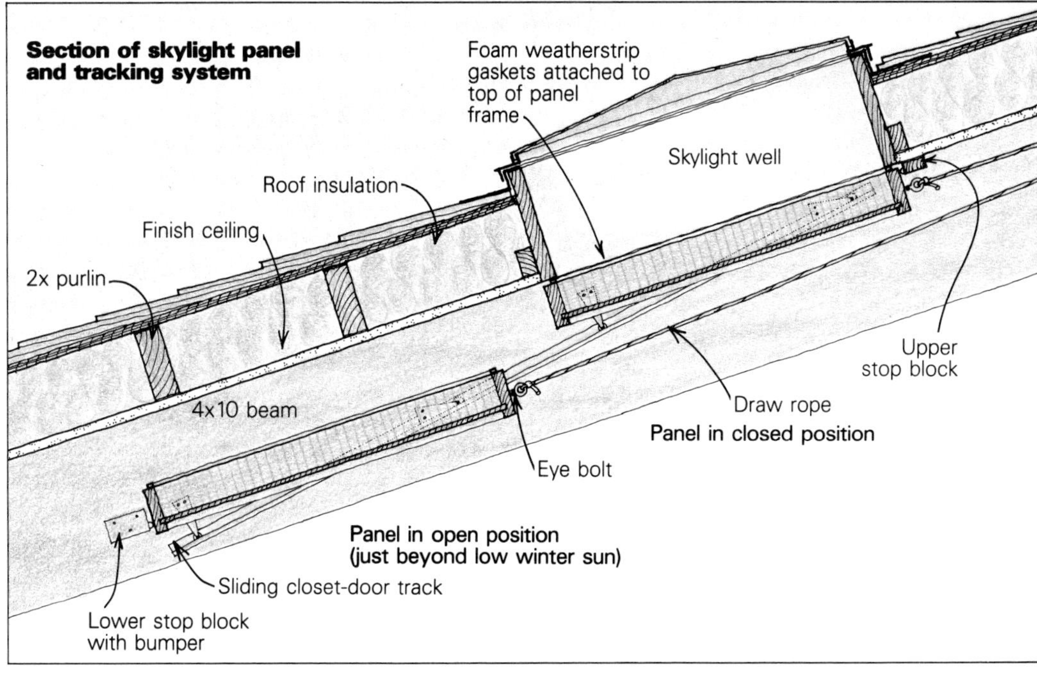

The insulated panels consist of a 1x3 fir frame that's joined with glued dadoes (detail photo at right) and grooved to accept a Masonite bottom. Inside the frame is a tight-fitting panel of 2-in. foil-faced rigid insulation board, with the foil side up. The panels ride in tracks (photo at top) cut from standard sliding closet-door hardware, and the wheels are mounted in a way that keeps the panel from making contact with the ceiling until it's directly under the skylight, where its gasket makes a positive seal. The rope-and-pulley system allows all the panels to be controlled from a single location.

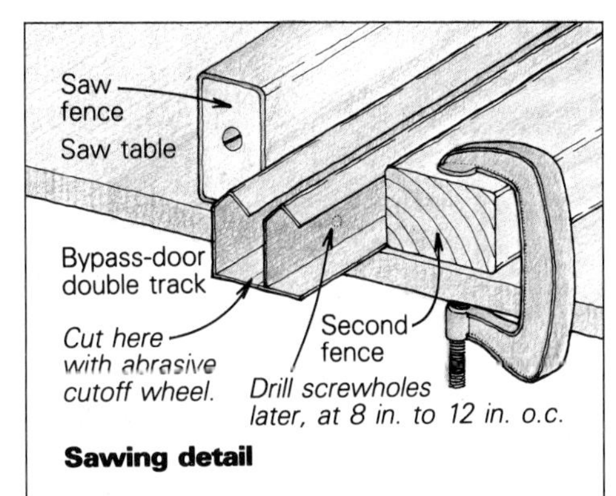

Photos above and facing page: Jonathan Landes; Drawings: Vince Babak

Counterweights with tubular-sock covers are sized to match the weight of the shutters. An easy tug on the cord and the shutters can be moved up, and they will stay put in any position.

side the panel can be any good-quality 2-in. foil-faced rigid foam. The foam should be fitted tightly into the frame. We use a panel adhesive that is compatible with the foam (check with the manufacturer of the foam) and glue it to the top of the Masonite to help eliminate warping. Make your panel frame a little deeper than the nominal 2-in. thickness of the insulation. You will often find that it is a little thicker.

Installation tips—Before mounting the tracks on the beam sides, screw the wheels on the panel frame. As the drawing on the previous page shows, they should be set at angles that will position the ends of the track within an inch of the top and bottom edges of the beam. The wheel-mounting angles are determined by the length of the panel and the depth of the beam. Laying this out schematically full scale on a plywood sheet can save a lot of tedious test-fitting.

Next, get a helper to support the panel against the ceiling in the closed position under the skylight well. The tracks can be slid into position over the wheels on each side of the panel and screwed in place. We use 1¼-in. drywall screws to fasten the tracks to the sides of the beams. Hold the track tightly against the bottom (the running surface) of the wheels until at least two screws are in place in each.

If we find that one or more of our beams has warped in curing, we make the panel to accommodate the smallest width in its run. Then we shim out the tracks in the wider portions of the run so they're exactly parallel to each other. For shims, we use either small precut pieces of stiff cardboard or thin washers sprayed flat black.

Once you have the exposed part of your tracks screwed to the beams, move the panel to that part of the track so that the rest of the track can be fastened. You may find upon moving the panel back and forth that you will want to adjust the shimming slightly. Sometimes this is just a matter of screw pressure.

When you are satisfied with your track operation, install two braking blocks on the downhill end of the run. These stop blocks should be of the same material as the panel frame and finished similarly. We also fit them with rubber bumper buttons, which are available in most hardware stores. If your panel is in the closed position on the downhill end of the track, adjust the stop blocks' position so the panel can come into full contact with the ceiling. The blocks' primary function is to stop the panel's downhill movement without damage, should someone let it down too fast.

Stop blocks should be installed at the upper end of the run, too. These will keep the panel from being drawn off the end of the track.

Tying things together—The draw rope is tied to an eyebolt set in the center of the frame facing the upward direction of the run. There is considerable stress placed on this piece and so when assembling the frames we glue the dadoed joints together, as shown in the photo at bottom right on the previous page.

When devising our rope layout, we gather the rope ends to the most convenient location for owner operation. The primary source of friction is in the pulleys used to make the rope turn corners. So buy good-quality pulleys and use the largest size that is practical for your design.

Once the pulleys and ropes are in place, all that's left to do is attach the counterweights to the ends of the ropes so that the panels can be raised and lowered without strain. We tie a temporary loop in the end of each rope and hook a small spring scale to each one in turn. When we pull on the scale, we find two weight levels indicated. The first level is the force needed to stop the downward progress of the panel on its track. The second level is the force needed to begin moving the panel up its track. The range between the two represents the amount of force needed to overcome the friction in the system. We size our counterweight to be within this range on each panel. This allows the panel to be set partially open if desired.

We use 1½-in. round steel stock for our weights, and cut it to the length that will give us the appropriate weight. Finally we drill a hole in the end for the rope and tie it up. As a finishing touch, we cover the counterweights with tubular socks (photo left).

Using the system—This panel design will give you approximately R-16 nighttime insulation over your skylights. It is simpler to devise a high R-value insulating cover for a skylight than it is for a vertical south-facing window because panel storage is not a problem for skylights. Convenience and ease of operation ensure that the owners will be more likely to use the system, and therefore that it will be more effective in saving energy.

In the summertime, the panels may be closed during the hottest part of the day. The foil covering on the insulation may be left exposed to reflect as much light as possible. If the panel tops are visible in their closed position from outside the house, a likely compromise would be to paint the upper foil surfaces white or cover them with light-colored fabric. The sides of the well should always be painted a light color for proper light transmission.

Some summertime heat, however, is bound to be trapped in the well and accumulate to a fairly high temperature. If you glaze your skylights with Plexiglas, you will want to protect it from temperature buildups. One way to do this is to have a seasonally adjusted blocking device to allow your panels to remain just slightly open so that a small amount of convection will moderate any high temperature buildups. Another solution is to build some simple, easily handled lattice boxes that are installed seasonally over the skylight and fastened with wing nuts to hanger bolts set in the skylight curb. With the lattice in place, one may still benefit from the daylighting effect on all but the warmest of days. □

Larry Medinger, of Ashland, Ore., designs and builds energy-efficient passive-solar houses.

Installing a Factory-Built Skylight

Careful selection and minor structural modifications make for a good job

by Jim Picton

Many home owners and craftsmen who are willing to cut into floors and walls will steer clear of retrofitting skylights because of their reputation for developing leaks. This notoriety is largely a result of the failure of older or improperly flashed units, which rely entirely on a chemical seal like asphalt roof cement to keep water out. One such variety, still available today for about $25, is the plastic bubble—a sheet of acrylic with a bulge in the middle—which is plopped down in a bed of cement, with shingles laid over the edges. Successive layers of cement are then applied to the edge surfaces as often as necessary.

We can thank the Arab oil embargo for improving the standards of skylight construction. Interest in alternative energy sources has brought solar heating to the fore, and with it the need for reliable roof windows. While there are still times and places when you will want to build your own units, competition in the marketplace has produced a number of well-designed skylights that are as weatherproof and problem-free as conventional vertical windows. You should still shop carefully, though. Prices vary widely, and are generally an indication of quality.

There are several things to look for when you're shopping for a unit. First, decide whether you want a fixed or operable skylight. Second, consider the flashing package. Step flashing will last the life of the roof. Strip flashing won't. Third, see if the skylight has a thermal break. On some, metal extends from the outside surface to the inside. This can cause a lot of condensation trouble in a cold climate. Fourth, check to see if you need tempered glass. You can save $50 or so by not using it, but building codes specify it where standard glass might easily be broken. Finally, examine the screen setup if you're buying an operable unit. Some are easy to use. Others are a bit quirky: For example, my Velux screen opens and closes with Velcro strips.

Most skylights, whether site-built or manufactured, have curbs 4 in. to 6 in. high, which raise the glazing above the level of the shingles and divert water around the unit. The glazing is spared the cascade that develops as water from the rest of the roof flows to the gutters. In addition, curbs keep granules from the shingles and other debris off the glass. This prevents scratching, and prolongs the life of the glazing seal.

A fixed-glass unit will run between $200 and about $600, depending on flashing and other

Picton's operable skylight was installed between reinforced rafters in a room with a ceiling that followed the roof pitch.

materials. You can get operable skylights for as little as $200, but most cost a great deal more. Crank, spring-loaded and center-pinned units are available, some with strip flashing and some with step flashing. With my contractor's discount, my center-pinned Velux and its flashing package cost me $500.

Structural considerations—If you have a flat ceiling with an attic above, you'll have to build a shaft between the roof and the ceiling. When roof and ceiling are separated only by rafters, as in the installation shown above, the job is more straightforward. In either case, before cutting a hole in your roof, think about the effect of cutting through one or more rafters. As a start, consider how roof openings are framed in new buildings. Double framing is conventional wherever rafters or headers are used to support additional roof-framing members. Rafters and headers that frame the opening are doubled. Cripple rafters run from the headers to the top and bottom of the roof.

Installing double headers is hardly more trouble than installing single ones, but doubling the rafters can be a problem. To be most effective, the double rafters should extend all the way to the points of support—usually to the ridge above and to the double top plate of the wall below.

You can usually do most of your cutting and framing from inside without disturbing the exterior roof surface. This lets you work without worrying about the weather. In some installations though, the whole point is to disturb the interior finish as little as possible, and you may decide to remove some of the sheathing from the outside of the roof, and install double rafters from above. This involves additional reroofing, but may save you the grief of having to live in the dust and rubble caused by tearing up the ceiling.

The added load taken by the rafters on each side of an opening can also be offset by reducing their span, or the distance between points of support. If there is a small attic crawl space above the opening, a purlin can be snugged into the area between the collar ties and the rafters. Although collar ties themselves reinforce the rafters, a purlin will cut down the span of the rafters and transfer the load to the walls.

Another alternative is to double up only a portion of the side rafters. A piece of lumber the same thickness and width as the rafter, extending a few feet above and below the skylight opening, and nailed solidly to the rafter, can have a stiffening effect.

When you're selecting a skylight, consider the width of the unit in relation to the rafter spacing. Choosing the next smaller size may mean cutting through one less rafter. Get the advice of a professional, a structural engineer or the local building inspector, if the problems you foresee do not have direct solutions.

Roughing out inside—Once the location for the skylight is selected, mark the rough opening for the unit on the ceiling. It should be about ¼ in. larger all around than the outside dimensions of the skylight unit you've settled on. Measure up 3 in. from the top line you've marked, and down 3 in. from the bottom one. You'll need this extra space for the new double headers. Next, find the rafters at either side of the opening and mark cut-lines that follow their centers. When the opening is completed, the ceiling will be patched at these lines, and the rafters will provide a nailing surface for drywall or finish trim.

Cut the ceiling using a utility knife, or a skillsaw set to cut only the thickness of the ceiling material. You don't want to cut through hidden wiring. When the ceiling panel is removed, pull out any exposed insulation and re-route the wiring if necessary.

To prevent the roof from sagging slightly when the rafters are cut through, support the rafters above and below the opening with temporary braces made by knocking together two 2x4s in the shape of a T, and wedging them between the rafters and ceiling joists. Then cut the rafters to be removed square, 3 in. back from the rough opening line.

Install the double headers one at a time. Cut the first piece to fit between the side rafters,

Picton removes a rafter, above, having already reinforced those on either side by doubling them up. The T-brace to his left keeps the roof from sagging before headers are installed.

Both boards of the double header are cut to fit between the side rafters, left, then installed one at a time. They are toenailed between the side rafters, then face-nailed. In this installation, the center nailing surface for the top header is the end of a collar tie.

Once the roof has been opened up, the skylight unit can be lifted into place, below. After it is checked for square, it can be fastened to the roof. A helper comes in handy for this part of the job.

and toenail it with 16d nails. Check to make sure you've got a square opening that allows about $\frac{1}{4}$ in. of clearance on all sides of the skylight unit. Then face-nail through the header into the end grain of the cripple rafters. Toenail the second header to the side rafters, and face-nail it to the first header. After this framing is complete, stuff any remaining openings with insulation, and then make the necessary repairs. If the ceiling is drywall, this is a good time to get a first coat of tape on the patch.

If you are contemplating re-shingling your roof, now is a good time to do it. If not, you will need to remove some of the shingles at the top and sides of the opening in order to flash the new skylight. This isn't a big problem, and it has the advantage that most of the shingles at the sides and bottom of the skylight will be cut and left in position, eliminating the need to mark and trim each shingle individually.

The roof opening should be cut from the outside. Locate the opening by driving nails up through the roof at the four corners of the opening you've just framed up inside. String chalklines and snap the perimeter on the shingles. Then pound the nails back through the roof, and remove them. Now you've got a chalked outline of the rough opening. The shingles should be cut back to about $\frac{3}{4}$ in. from the edges of the unit, so you need to mark a cut-line $\frac{1}{2}$ in. outside the line you've snapped. Double-check the measurements of both the skylight and the rough opening.

Asphalt shingles are easy to cut with a circular saw and an old carbide-tipped blade. The carbide tips can be dull or chipped, as long as the teeth are widely spaced. Asphalt material quickly gums up ordinary sawblades and makes them useless. Set your blade depth to avoid cutting into the sheathing, then cut the shingles along the outside lines. Remove and discard them. Now mark the rough opening on the sheathing, using as guides the holes left by the corner nails. Cut the roof sheathing out along these lines with a better, sharper blade.

Installing the unit—At this point, the skylight can be set in place. If you're not re-shingling, you'll have to lift some shingles from around the edges of the opening or remove them by popping nails with a flat bar, so that you can mount brackets or install flashing, depending on the design of the skylight. When the unit is centered in the opening, check it for square and be sure it operates correctly. Then attach it securely according to the manufacturer's instructions or the design specifications.

If you are installing a new roof, the new shingles should be applied up to the course whose lower edge is within 10 in. of the opening. The bottom flashing for the skylight can then be installed.

Most skylight manufacturers offer a flashing package with their units. It's usually more expensive than flashing you can make yourself, but if you have a problem with the skylight, the manufacturer could void your warranty if you haven't used his flashing, even if you've done a good job with your own. If flashing has been provided by the skylight manufacturer, install it

Flashing is provided with prefabricated skylights, or can be purchased. Above, the first piece of step flashing is actually an extension of the bottom flashing. It's trimmed so it won't extend below the side shingles, and fastened in place with a nail in its upper outside corner—the proper technique for all step flashing.

With the side shingles trimmed and the top two courses along the skylight's head removed, the top flashing is set in place, right. The top flashing supplied with this unit has a bent return along its upper edge to keep backed-up water from getting under the roofing.

Full shingles are used as the starter course at the top flashing, below right. Like the shingles that will overlay them, they are worked up under those already on the roof, and are nailed high enough so the flashing won't be punctured.

according to the instructions provided. If the instructions merely tell you to nail the flashing to the roof, do so with two nails only, at the upper outside corners of each end.

Some manufacturers supply decent base and head flashing, but rely on strip flashing and mastic along the units' sides. I have little faith in strip flashings, and if that was all the kit included, I would probably make and install my own step flashing, as explained on pp. 108-109.

The inside surfaces of the framed skylight opening can be trimmed with wood or covered with drywall. Depending on the design of the skylight, drywall butting into the bottom of the curb may require a small wood molding to finish it off, or you may decide to flat-tape this gap or rabbet the curb to accept the thicknesses of the finish material.

Your pleasure in looking at the installed skylight for the first time is balanced by your anxiety during the first good rain. No leaks are likely in a proper installation, but you might see wet corners or dripping glass under certain conditions. Condensation is a problem not limited to skylights, and humidity and cold can turn any large expanses of glass into a marvelous condenser. A number of methods exist for controlling humidity in a house, and skylights require no special treatment. If you notice dampness around the skylight, check its source. If it's condensation, forget it. □

Jim Picton is a carpenter and contractor in Washington Depot, Conn.

Site-Built, Fixed-Glass Skylights

An energy-efficient, watertight design that you can build with standard materials

by Stephen Lasar

Fixed-glass, site-built skylights can provide many cost and thermal benefits for new construction, as well as for additions to existing structures. And having a reliable design that can be built on site gives the architect or builder the flexibility to meet functional and aesthetic requirements. The skylight design that we use most often is based on standard techniques and materials, and any skilled carpenter can install them. So far, our skylights have withstood a wide range of weather conditions without failing. And the prepainted aluminum flashing and battens we use give these skylights a clean, unobtrusive appearance.

The curb—In holding the glass above roof level, the curb keeps runoff and debris off the skylight. The curb is made from straight-grain 3x10 Douglas fir that we rip in half to yield two 3-in. by 4½-in. curb members. We then rabbet them along one edge to accept the glass. Center curbs, if any, hold two glass panels, and are rabbeted along two edges. The depth of the rabbet is 1¼ in., which is ¼ in. less than the combined thickness of the insulated-glass panel with its two glazing strips. We take up this extra ¼ in. by compression when we screw down the battens to form a weathertight seal (drawing, facing page). The rabbet is ¾ in. wide. We size the curb enclosures so that ½ in. of this width holds the glass and glazing tape. This leaves ¼ in. of open float space between the edges of the panel and the wood.

We assemble the curb box on the roof, using butt joints at corners and a housing (dado) to let each center curb into its top and bottom curb. Toenailing the curb to rafters and headers is usually enough to hold it securely, but if the roof pitch is steeper than 12 in 12, then we use metal clips, too. For structural reasons, we use 3x10 or 3x12 rafters on most spans beneath central curbs (photo below left).

Site-built skylights call for flashing that is also cut and formed on site. The system shown on p. 109 works well with this type of curb. The aluminum should be .019 in. or thicker, and at least 8 in. wide. I always specify prepainted flashing stock because it blends in with a new roof. Shiny, unpainted aluminum draws needless attention to the skylight.

Tape and glass—The inside tape we use is Pre-shim 440 ¼-in. by ½-in. spacer-rod tape made by Tremco, Inc. (10701 Shaker Blvd., Cleveland, Ohio 44104). It's a butyl-base material and is compatible with the silicone second seal of the insulated-glass units we use (be sure to check caulk-sealant compatibility very carefully). The spacer rod, or shim, is continuous, and limits the compressibility of the tape, so that the glass won't settle away from its seal over time. It also helps to prevent tape squeeze-out at pressure points. Sticky on both sides, the tape comes in rolls, with one side faced with paper. It is applied to the bottom of the curb's rabbet, set slightly back from the vertical edge. It's important to lay the tape carefully, so that it has room to expand without touching the inside of the rabbet or protruding over its outer edge. Butt each tape so there are no airspaces between sections, and lay the tape with the paper facing up (photo below). Don't strip off the paper yet, because you may have to shift the glass panel after you lay it down.

The glass is installed from the outside, so be sure you rig a secure scaffold. The insulated panels used here are standard, double-pane tempered sliding-door units, 92 in. long, 34 in. wide and 1 in. thick. Don't try to lift and position the glass by yourself; get at least one helper. You won't need suction cups for getting the glass onto the roof, but you will need them to lower the panels down into the curb and to shift them into final position.

Before setting the glazing in place, put a neoprene setting block (¼ in. thick, 4 in. long and 1 in. wide) one-quarter of the way in from each bottom corner of the rabbeted curb. These spacers hold the panel away from the bottom edge, giving the glass room for expansion. Now you can set the bottom edge of the panel against the setting blocks and lower it carefully into the curb. Center the panel and make cer-

With the rabbeted curbs nailed to rafters and flashed to the roof, the next step, above, is to lay the glazing tape. The paper should be left on the tape until the glass panels are centered. Then the top layer of glazing tape is applied, right. A batten joint will hide the flashing and tape that cover the curb between panels. Tape is laid against the edge of the glass to allow ¼ in. of float space on all sides.

From *Fine Homebuilding* magazine (October 1982) 11:62-65

tain that you've got your ¼ in. of float space around all four edges. To do this, you'll have to pull up the panel at the top, adjust it, and set it down again. Once the position is right, pull up the panel one last time, strip the paper facing off the glazing tape and set the glass panel down for good.

The next step is to apply the outer glazing tape. We use Tremco Polyshim tape, a butyl-base, compression-type tape with good adhesion and elastic qualities. As with the tape beneath the glass, this glazing strip shouldn't butt right against the wood. Leave about ¼ in. of space between wood and tape for expansion. Once you've taped the glass, you should also flash and tape all corners of the curb as shown in the small photo on the facing page. Wherever aluminum battens will intersect, cover the curb with a strip of flashing and a length of tape. Remove the shim from the tape so that when the battens are screwed down, the glazing tape will be squeezed into the joint to form a weathertight seal.

For battens over the curb's perimeter we use 3-in. by 3-in. by ⅛-in. pre-painted aluminum angles. We cut the battens for center curbs from flat aluminum stock 3 in. wide and ¼ in. thick. The side angles should be cut with ears at both ends to create an interlocking joint with top and bottom battens, as shown in the drawing at right. All top and side pieces also need to

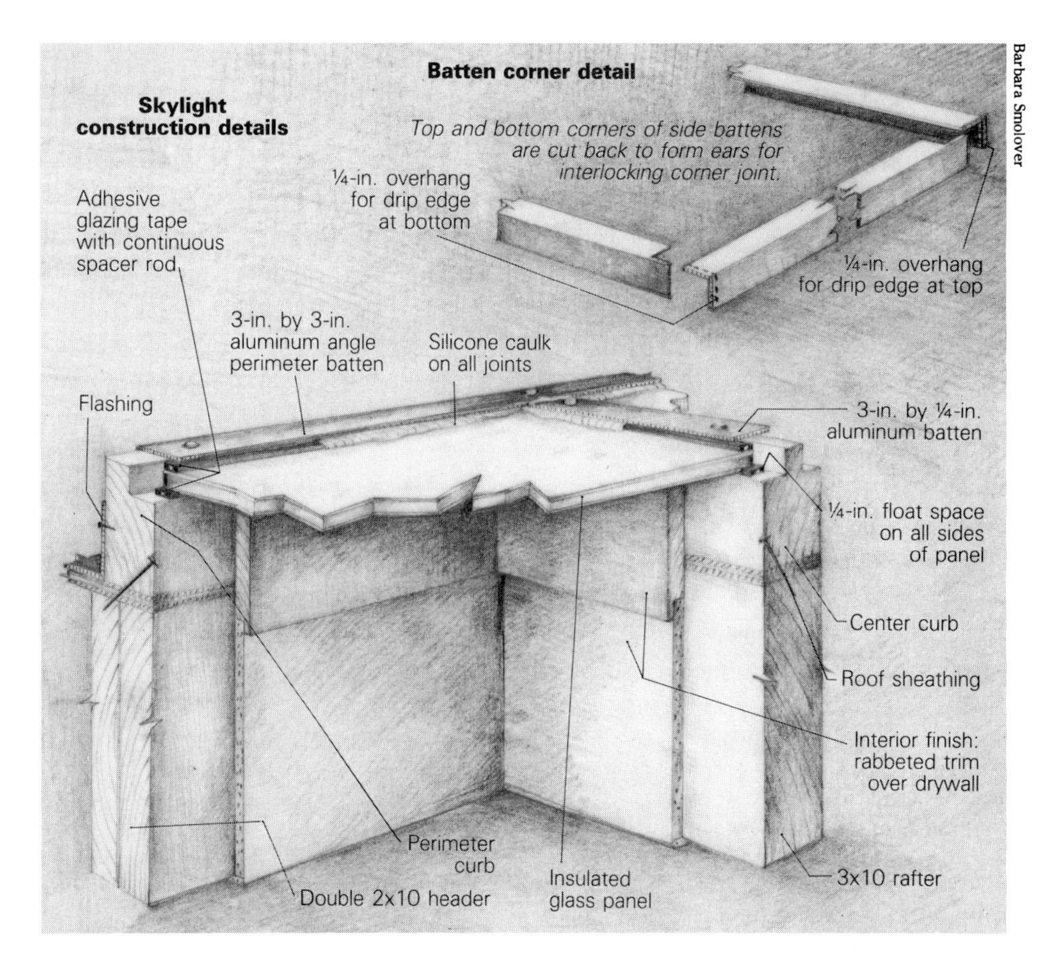

Barbara Smolover

Below, a flat aluminum batten is laid down over a central curb. It has been predrilled for screws, and will cover the top layer of glazing tape at the edge of the glass panel. Details of the skylight construction are shown in the drawing above.

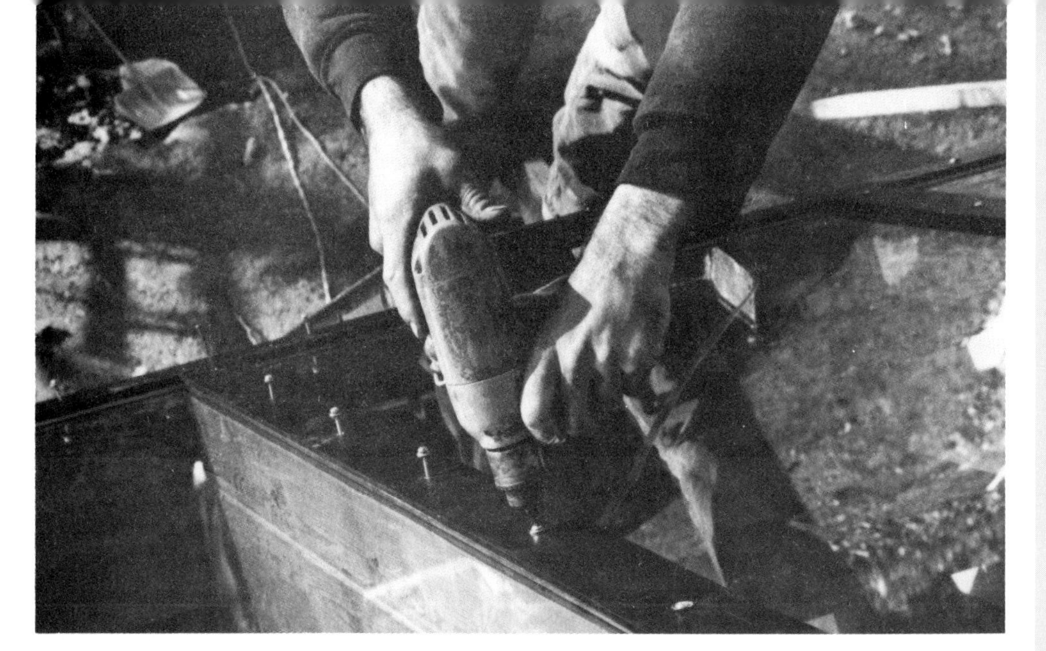

Top, stainless-steel screws with neoprene and stainless-steel washers pull the battens tight, compressing the glazing tape for a weathertight seal. Above, the last step in building the skylight is to caulk all batten and glass-to-batten joints. For best results, warm the caulk before you start to apply it and keep the bead continuous.

be cut slightly long to create drip edges. The perimeter battens cover the curb flashing. Install them first, over a generous bead of silicone caulk. Where battens meet, leave about $\frac{1}{16}$ in. between metal edges, for caulking and for the glazing tape to squeeze into. We use $1\frac{1}{2}$-in. long stainless-steel No. 10 Phillips panhead screws with stainless-steel and neoprene washers to pull the battens down tight over the glazing. As for screw spacing, 16 in. o.c. is good for angled battens; 12 in. o.c. for flats.

Replacing a defective panel of insulated glass is an irritating and expensive job. Not considering breakage, failure of the seal or spacer between panes is the most common problem with insulated-glass panels. It's a good idea to specify a double seal that has a good rating from the Insulating Glass Certification Council. This group publishes "Certified Product Directory," a 28-page booklet that lists the names, addresses and phone numbers of over 50 American manufacturers of insulated glass, along with details on their products' corner construction, spacer materials, dessicants and sealants. Though intended mainly for manufacturers, it may be of use to architects, contractors and builders. It costs $1 from the IGCC (Industrial Park, Route 11, Cortland, N.Y. 13045).

Caulking is the last step. The joint between the glass and the aluminum batten and the joints between battens should be sealed with an even, continuous bead of exterior-grade silicone, carefully lapped at corners. For best application, get the caulk warm before you start using it. Fill any voids or uneven spaces, and smooth the caulk with your finger where necessary.

There are several options for finish trim on the interior of the skylight. We use either veneered plywood nailed directly to rafters and headers, or a trim strip, rabbeted to cover plaster or drywall. In either case, it's important not to butt the plywood or trim against the glass surface. Allow $\frac{1}{4}$ in. to $\frac{3}{8}$ in. all the way around the skylight so the glass and wood have room to move. □

Solar architect Stephen Lasar practices in New Milford, Conn.

Flashing a curb

by Jim Picton

Most manufactured skylights can be bought with flashing kits, but if you're installing a site-built unit like the one described in the previous article, cutting and forming your own flashing is the only way to go. Even if you're working with a factory-built unit, you may choose not to buy the optional flashing kit, but to make your own instead. This is an especially attractive alternative when the manufacturer supplies strip flashing for the unit's sides instead of the superior step flashing. Buy a roll of .019-in. thick aluminum 16 in. or 20 in. wide, and follow the directions below.

Base flashing—**First unroll a length of aluminum, and cut off a piece about 8 in. longer than the width of the skylight. Slice it to a width of 8 in., and save the scrap for step flashing. Bend the aluminum lengthwise along a straightedge to form a right angle, with one side about 3 in. across and the other, 5 in. Then set the 3-in. side against the bottom curb of the skylight, and the 5-in. side flat on the roof. Hold the angled aluminum so that an equal amount of flashing extends beyond each side of the skylight. Now mark a line on the vertical side of the flashing flush with the corner of the curb on each side of the skylight. This is a fold line. Next, draw a line to make a diagonal cut across the vertical side of the flashing, ending exactly at the point where the fold line meets the bent corner of the flashing, as shown in step 1 of the drawing on the facing page.**

When you've cut both sides, fold the vertical half of the flashing around the skylight, and flatten the rest onto the roof. If you're not re-shingling, slide the angled tab of flashing under the first course of shingles on each side of the skylight. This is your base flashing. Folded and cut, it's ready to tack to the roof with a single nail in each upper corner.

Step flashing—**Most manufactured flashings include the first two pieces of step flashing as an integral part of the bottom flashing. Each piece is bonded to the bottom flashing with a soldered or locked seam joint. You can approximate this detail using an 8-in. square of aluminum sheet from your roll, bending it at a right angle with 3 in. vertical and 5 in. flat, and making a V-cutout along its fold so that it can be bent around the base (step 2).**

The rest of the step flashing, except for the two top corner pieces, consists of square or rectangular sheets with a single 90° bend. The precut sheets sold in hardware stores as step flashing are typically 5 in. by 7 in., a size that offers minimal protection. Buy larger pieces if you can, or make your own from 8-in. squares.

If you're putting on a new roof, lay the next course of full shingles when the first piece of step flashing is in place. Then mark the shingle that will fit next to the skylight and cut it, leaving about $\frac{3}{4}$ in. of space between curb and shingle edge. Position the shingle and nail it down at the top edge farthest from the skylight. Now slide the second piece of step flashing under the shingle adjacent to the skylight curb, so that the bottom edge of the flashing is about $\frac{1}{2}$ in. higher than the bottom edge of the shingle. The flashing should lie directly under the bottom half of the upper shingle, and over the entire concealed upper

half of the lower shingle (step 3). If you're step-flashing a roof that's already been shingled, be sure to lace each piece of flashing over and under the shingles before nailing its upper corner to the roof.

Head flashing—The last pieces of step flashing will be cut and bent around the top corners of the curb (step 4), just like the step flashing at the lower corners. Once you've done this, the head flashing can go on. Bend, cut and fit it just as you did the base flashing, but use a full-width piece (at least 16 in.) from your roll. Before nailing it to the roof, nail one thickness of shingles to the bare roof above the curb. This will shim the roof up to the level of the shingles on each side, and prevent the head flashing from sagging in the middle and collecting water and debris. Remember that the upper part of the sheet must go under the first course of shingles above the curb, while the edges rest on top of the corner step flashing (step 5).

An exception to this procedure is when you are using a manufactured skylight such as Roto Stella, which requires installing the top and bottom flashings at the time the unit is installed. In this case, you have to slide the last piece of step flashing up under the preformed corner of the top flashing, and forget about bending the step flashing around the top.

Manufactured flashings usually won't extend too far up under the shingles, so they often have a bent return on the upper edge of the top flashing. In case water manages to back up under the shingles, this feature is intended to prevent the moisture from traveling beyond the top of the flashing and getting back under the roof.

Site-constructed flashings that are run 16 in. or 17 in. under the shingles do not need a bent return, but you can make one as an added precaution. Using a straightedge and your fingers, bend the top of the flashing up, about ½ in. from the edge. Fold it all the way over, then place a board over it and hammer it flat. Finally, use a screwdriver to pry the bent edge back up a little, so a space is visible between the top edge of the bent return and the rest of the flashing below.

The most difficult part of fitting homemade flashing is working around the four corners of the skylight curb. The points of these corners are the one area in a flashing job where, unless you are using soldered, welded or otherwise pre-formed components, no overlap is possible. The way to solve this problem is to do what carpenters and roofers always do when faced with the impossible: caulk it.

Counterflashing—Sometimes called angled or perimeter battens (drawing, p. 107), counterflashing is a piece of metal (usually a heavier gauge than roof flashing) bent lengthwise at right angles. Half of this flashing covers the top of the curb; the other half protects the sides of the curb and covers the top edges of the roof flashing. On some prefabricated skylights, counterflashing is permanently affixed to the curb, and the step flashing has to be slipped under and worked into position as the roof is shingled. Other units have a removable counterflashing that is attached quite simply with screws once the roof flashing is complete. In either case, it completes the exterior seal. □

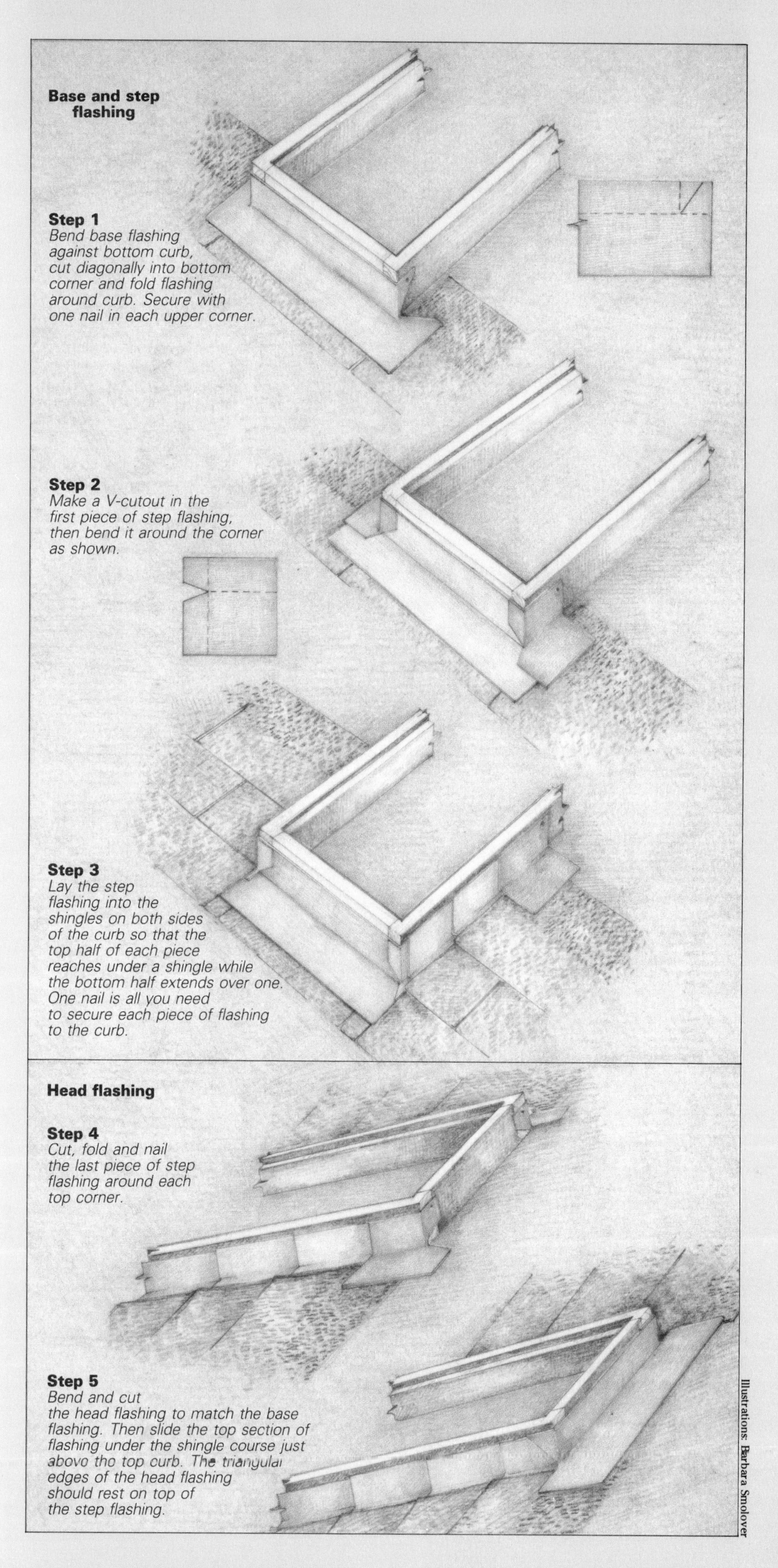

Curbless Skylights

Insulated glass mounted flush with the roof in a low-cost, site-built design

by Rob Thallon

Builders find themselves installing lots of skylights these days. There are some excellent manufactured skylights on the market, the best of which are well insulated and can be opened for ventilation and cleaning. These top-of-the-line units are expensive—$25 per sq. ft. or so. Less costly store-bought skylights are not operable, usually not insulated, and they are often translucent rather than transparent. Some of these budget skylights come with an attached, pre-flashed curb to raise the glazing above the roof, but many require on-site curb construction.

The curbless skylights that I started to build about nine years ago are installed more like large shingles, in contrast to most of the manufactured or site-built skylights that I've seen. As the photo at right shows, eliminating the curb gives the skylight a lower profile so that it looks more continuous with the roof.

In the last six years my partner, David Edrington, and I have installed more than 50 curbless skylights, and they've held up well. We use insulated, tempered-glass panels in standard sizes when possible, and we've simplified construction details to the point that we now feel confident that our skylights are the best fixed skylights available.

Our design can accommodate just about any type of glass or acrylic panel (providing you allow extra room for expansion), and replacing the glass is an easy job. The design doesn't rely on caulks and sealants, which have unpredictable lives, but rather on the behavior of water in contact with metal and glass. With slightly different flashing details, these skylights can be ganged to form continuous bands of roof glass such as those found in greenhouses. And perhaps best of all, our skylights cost about $4 per sq. ft., if you use off-the-shelf sizes of insulated glass.

General suggestions—The glass and side flashing drains in our system are at a slightly lower angle than the roof, so the amount of pitch is important. We've used this detail in 4-in-12 roofs, but I don't recommend going any shallower than this.

The skylight construction details can be adjusted to work with most roofing materials. We try to use those that fit closely, like cedar or composition shingles. Roll roofing or metal would also work fine, but with looser-fitting materials like shakes or tiles, the step and side flashing dimensions should be increased.

Organizing materials—Since the size of the glass determines the size of everything else, this is the first part of the skylight to consider. Insulated tempered-glass door blanks come in a standard 76-in. length and in three standard widths—28 in., 34 in., and 46 in. We usually use the 34-in. wide blank, and the installation I'll describe uses this size.

You can also adapt these instructions for glass of any size. We've had smaller tempered windows custom-built, but they end up costing more than the larger ready-made ones. On the other hand, glass larger than 34 in. by 76 in. usually isn't strong enough to be safe. Consult your glazing supplier about the panel strength in relation to size and snow loads in your area, and check with your building inspector for the minimum glass-thickness requirements for skylights.

It's important to understand that ordinary window glass isn't recommended for skylights because of its relatively low strength and because, if it should break, a large piece of it falling into a living space could be extremely dangerous. The three types of glass recommended for skylights—tempered glass, safety glass and wire glass—have overcome this problem in different ways. When it breaks, tempered glass is supposed to dice into tiny bits, each no larger than 3⁄16 in. Safety glass (also called laminated glass) is a sandwich of two layers of ordinary glass held together by a layer of plastic. Wire glass has a network of tiny wires running through it that prevent the glass from breaking into dangerous shards.

Of all the kinds of glass, we prefer the tempered for skylights because it's the most transparent and, under uniform conditions, its strength exceeds that of laminated or wire glass by a factor of about four. The chief disadvantage of tempered glass is the potential for incomplete dicing when it breaks (for more on the dangers of tempered glass, see *FHB* #16, p. 21). We've never had such a problem with any of our skylights, but the possibility has emerged, so codes governing tempered-glass skylights have been stiffened here in Oregon. Building departments require insulated units to have a safety-glass layer on the inside or a screen below tempered glass. The screen has to be at least 12 USA-gauge wire with a mesh no larger than 1 in.

Once you've decided on the type and size of the glass, you can order the flashing. General dimensions for each flashing configuration are noted in the bottom drawings on the facing page; the dimensions in parentheses are for a 34-in. by 76-in. unit. The bends are straightforward and should be easy work for any reputable sheet-metal shop. We usually specify 26-ga. galvanized steel or 16-oz. copper, but prepainted or stainless-steel flashing will work fine. We don't use aluminum because it won't bend to these shapes without fracturing.

At our local shop, the flashing package for a 34-in. by 76-in. unit costs $32.75 for galvanized steel, $81.00 for copper, $42.50 for prepainted and $77.00 for stainless steel.

Roof framing and flashing—The rough opening for the skylight needs to be ½ in. wider (across the roof) and 2½ in. shorter (parallel to the rafters) than the dimensions of the glass. For example, the rough opening for a 34-in. by 76-in. glass panel should be 34½ in. by 73½ in.

Sheathe the roof to the edges of the rough opening (but don't let the sheathing project into the opening), and install the roofing material up to the bottom edge of the opening. Next, fasten 1x4s to each side of the rough opening so their top edges project a couple of inches above the sheathing. These are just temporary fences for the step flashing, but

Architect and builder Rob Thallon lives in Eugene, Oregon. Photos by the author.

From *Fine Homebuilding* magazine (December 1983) 18:36-39

Framing and flashing section at bottom

Clips screwed to the rough-opening framing cradle the glass from below.

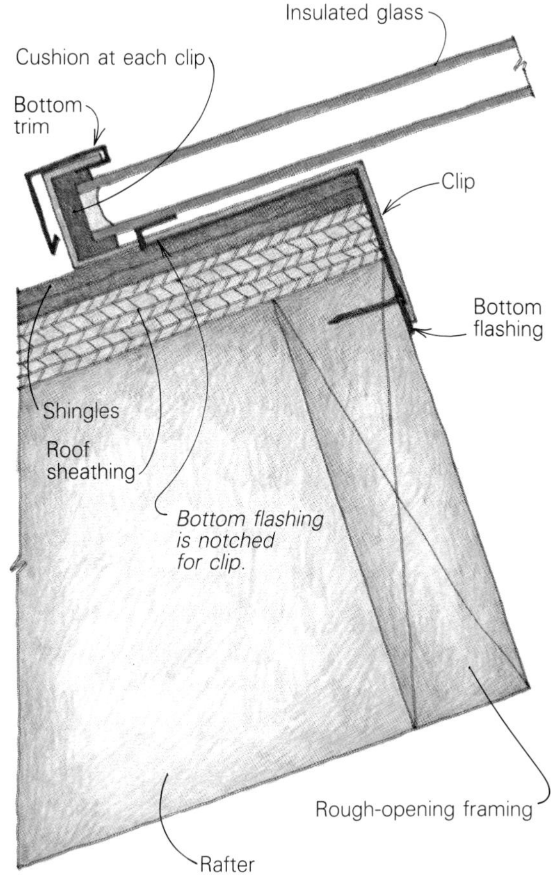

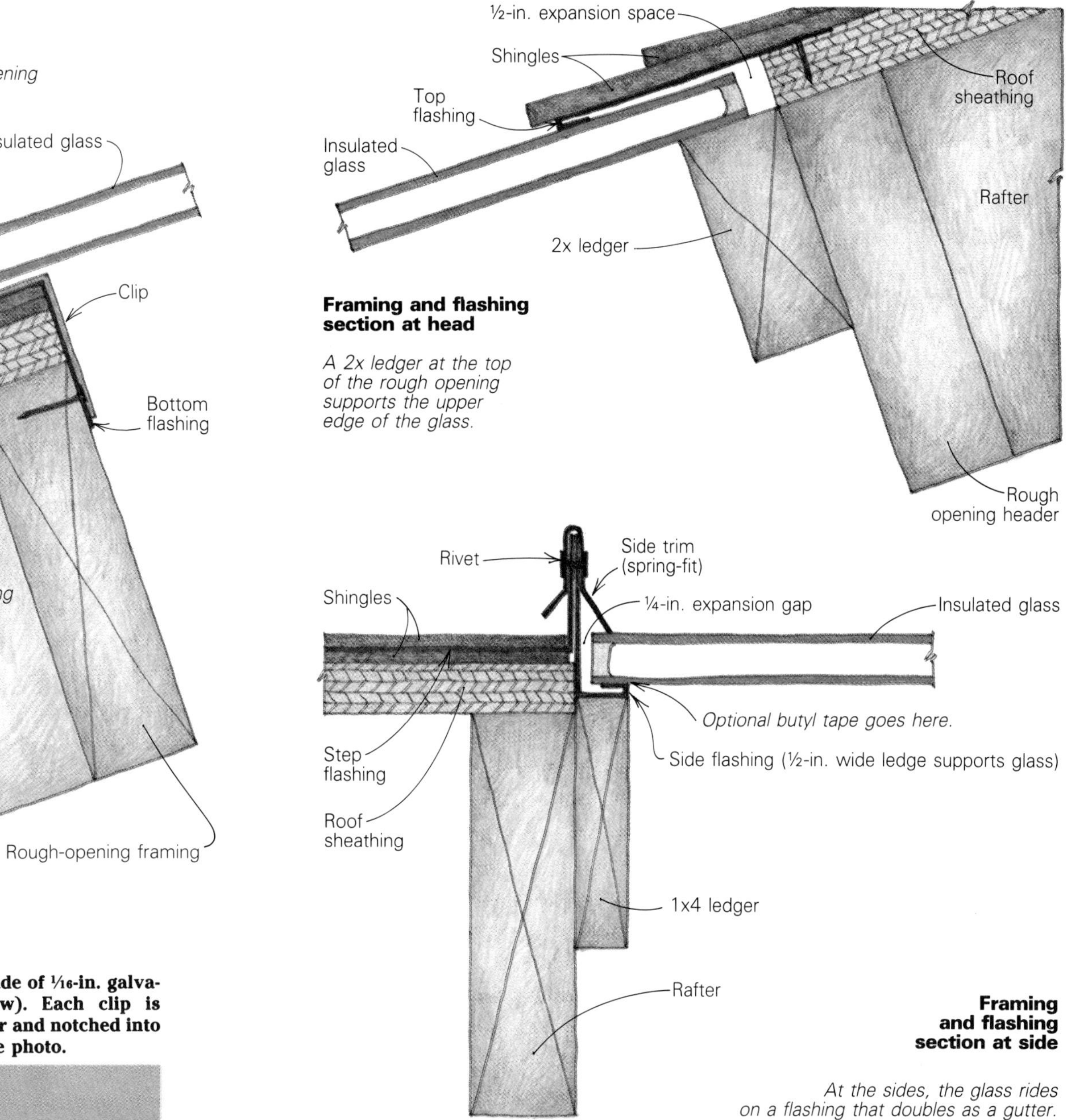

Framing and flashing section at head

A 2x ledger at the top of the rough opening supports the upper edge of the glass.

Framing and flashing section at side

At the sides, the glass rides on a flashing that doubles as a gutter.

The glass is held by clips made of ¹⁄₁₆-in. galvanized steel (drawing, below). Each clip is screwed to the bottom header and notched into the flashing, as shown in the photo.

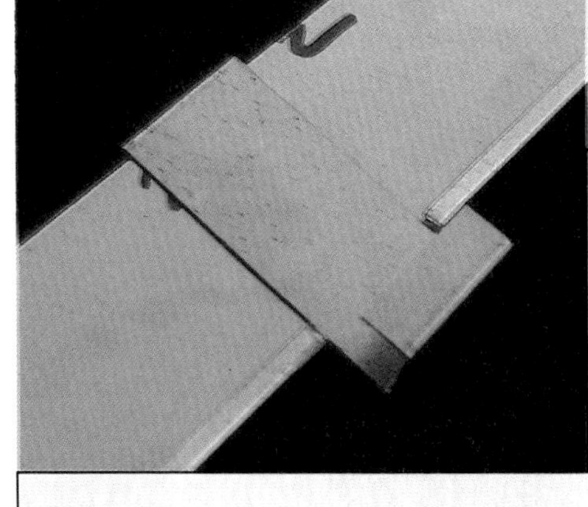

Flashing sections

All dimensions are for a 34-in. by 76-in. glass.

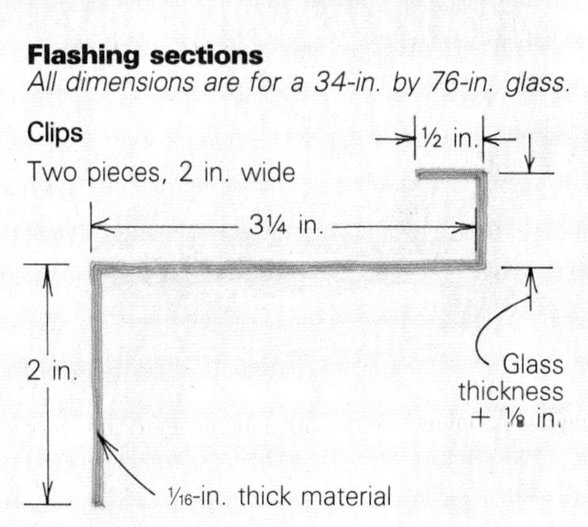

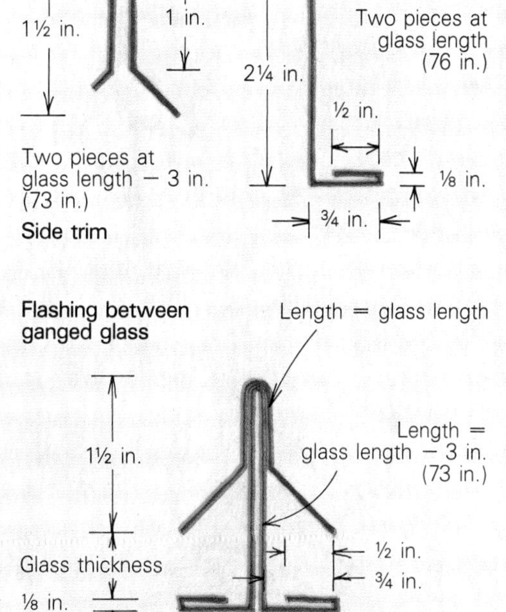

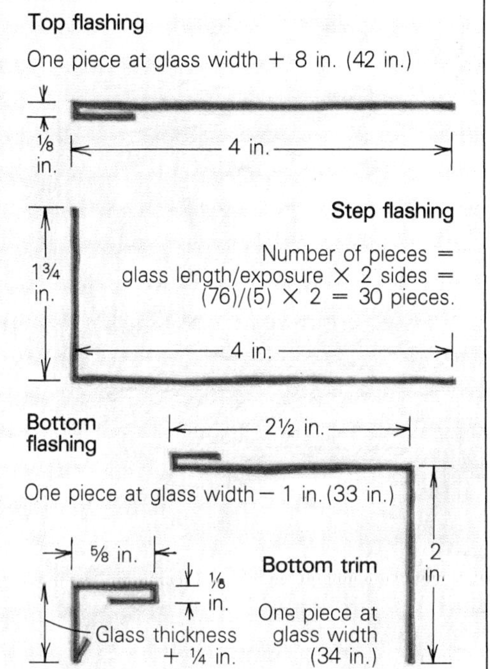

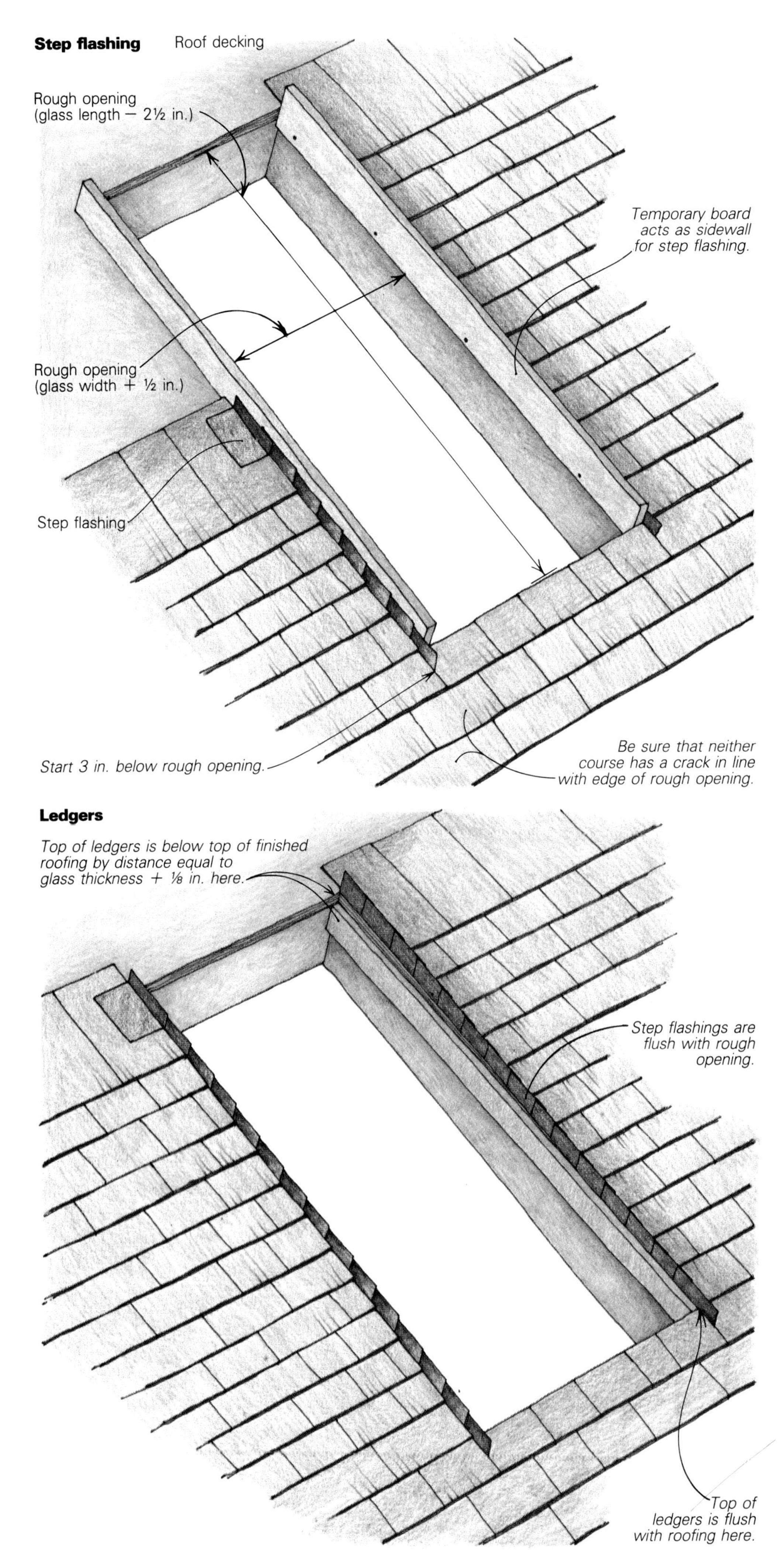

they should be fairly secure just the same, since you'll use them as though they were sidewalls. Continue roofing up the sides of the opening, installing the step flashing tight against the 1x4s (drawing top left). When the step flashing reaches the top of the rough opening, remove the temporary sidewalls.

Next, nail permanent 1x4 ledgers to the sides of the rough opening (drawing, bottom left). These should be cut from dry wood that has a moisture content of less than 15%, to reduce any chance of warpage. At the downhill end of the opening, make these side ledgers flush with the installed roofing. At the uphill end, set the ledgers below the surface of the roofing by a distance equal to the thickness of the glass plus ⅛ in. This is very important. These ledgers can be used as finish trim inside the opening, so you might want to use clear material and rip these pieces so they're flush with the ceiling finish.

At the top of the rough opening, install a 2x ledger with its top edge parallel to the angle of the roof, and set ⅛ in. above the side ledgers already installed. This detail is shown on the facing page, drawing top right.

Now you can install the flashing for the glass. The side flashing butts against the already installed step flashing, with its uphill end at the top of the rough opening and its bottom projecting 3 in. over the new roofing. The side flashing has a J-shaped profile, which creates a small gutter between the top of the side ledger strips and the sides of the glass panel. This gutter acts as a runway for moisture that gets past the counterflashing.

The side flashings don't need to be nailed to the ledgers to hold them in place, but if the roof is especially steep you might want to tack each one down at the very top to hold it steady during assembly. Use tin snips if necessary to trim the top of the side flashing flush with the top edge of the step flashing.

Next comes the only tricky part of the operation. Using tin snips, cut a notch out of the step flashing 3 in. down from the top of the rough opening and flush with the roofing, as shown in the top drawing on the facing page. Make a vertical cut in the side flashing 2¾ in. down from the top of the rough opening and fold the resulting flap onto the adjacent step flashing. There is now a ¼-in. wide tab in the side flashing that should be folded around the step flashing for a mechanical connection.

The tiny gap between the step and side flashings needs to be caulked or soldered against the weather. Soldering, which works on copper and galvanized, is more permanent than caulking, but also more difficult, so we usually seal this tiny crack with a gutterseal caulk made for galvanized gutters.

The bottom flashing is next. Cut out two notches in the bottom flashing wide enough for the panel clips to pass through and about 6 in. in from the bottom corners of the panel. Lay this flashing across the bottom edge of the rough opening and install the clips in their notches by screwing them to the header at the bottom of the rough opening (photo, previous page). Now you can install the glass.

Placing the glass—The glass panel is laid so that its edges rest on the side flashing and its bottom edge is supported by the clips. Neoprene cushions (setting blocks) inside the clips will lessen the chance of the clips starting a crack in the glass. We usually use the pads that protect the glass as it comes from the factory for this purpose.

As you get ready to install the glass, remember that a standard 34-in. by 76-in. double wall unit weighs about 105 lb. Be sure your footing is solid, and have as many hands available as is practical.

Here in Oregon, we haven't had problems with excessive air infiltration, but people in cold climates might want to bed the glass on a perimeter of butyl tape. Use narrow (¼-in.) tape to avoid clogging any of the drainage channels, and don't let the butyl tape come in contact with the edge seals of the glass panel—they aren't compatible and may cause each other to deteriorate.

With the glass in place, you can install the top flashing. (For installation details of the top, side and bottom flashing and counterflashing, see the drawings at the top of p. 111.) The top flashing fits 3 in. over the top of the glass, with its bottom edge resting at the notches in the side flashing. Lay this top flashing in place and tack it to the roof sheathing with a couple of nails near its top edge. Now finish the roofing, and be sure not to put any nails into the part of the glass that's hidden under the flashing.

The last step is installing the counterflashing. First slide the bottom trim piece over the clips so that it covers the bottom edge of the glass and is supported by the clips. This shields the sealant at the panel's bottom edge from the ultraviolet rays of the sun. Next slip the counterflashing pieces over the lip formed by the step flashing and the side flashing. This counterflashing is spring-fit against the glass. You'll have to cut out a small notch at the bottom of the counterflashing where it passes over the bottom trim. Fasten step flashing, side flashing and counterflashing together with three pop rivets per side—about 3 ft. o. c. If you're using galvanized flashing, dab caulk on these rivets to prevent rust.

If you want to make a skylight larger than the size of a standard tempered-glass unit, your best bet is to gang several together. We've done this often in solariums and greenhouses. The details remain the same except between adjacent pieces of glass. The flashing for this condition is a twin piece of side flashing (see p. 111, drawing bottom center). It's treated at the top and the bottom exactly like the side flashing already described. If, as is often the case in greenhouses, you want the glass to come right to the eave and drain directly into the gutter, you can eliminate the side ledgers, and just use the rafters to support the glass and flashings. This works especially well when the thickness of the roof sheathing approximates the thickness of the glass, as it often does. Here the glass can lie directly on the rafters, which can be finished to be an integral part of the installation. □

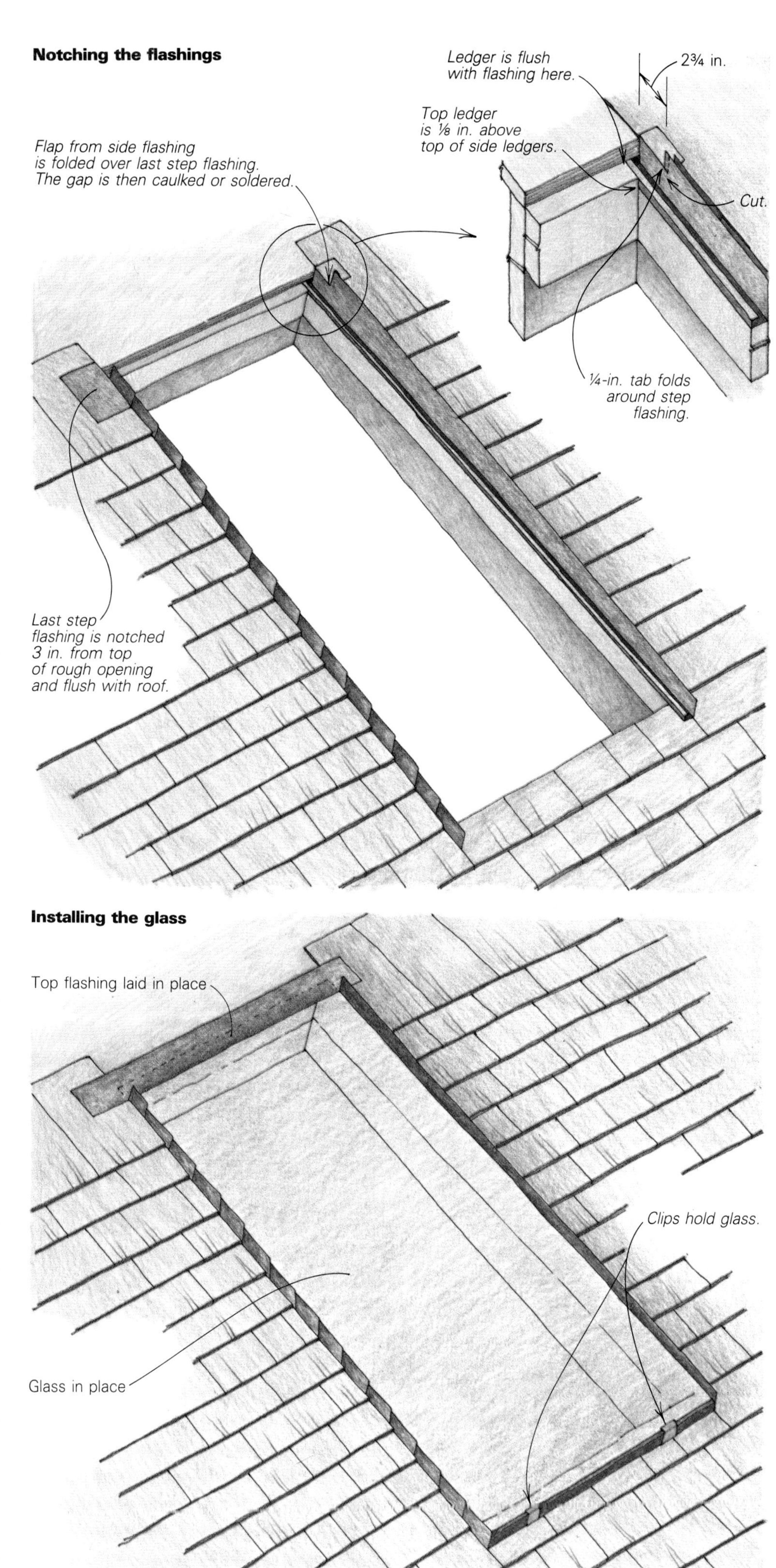

Acrylic Glazing

How and where to use this plastic instead of glass

by Elizabeth Holland

Commonplace now as the stuff of automobile lights, bank security windows, gas-station signs, camera and contact lenses, TV screens, and even paint, blankets and carpets, acrylic plastic has been around a long time. Although development of this highly elastic substance began back in the 19th century, it wasn't until the 1930s that chemical firms first began producing commercial quantities of acrylic, which can be manufactured as a liquid, as fibers or in sheets. And it took World War II, when the War Department started testing and using acrylic extensively in aircraft, to push the technology into the applications familiar to us today.

The larger family of plastic glazing materials has been closely scrutinized over the last decade by solar designers and builders searching for the least expensive material for collectors, greenhouses, windows, skylights and water storage. Most plastic glazings are flexible, lightweight, impact-resistant and light-diffusing. Acrylic stands out because it will not degrade or yellow in ultraviolet light. Along with high clarity and an impact resistance of 15 to 30 times greater than that of glass, acrylic offers a lifetime gauged at 20 years.

Given a burning rate of Class II in the codes, acrylic burns very rapidly, but does not smoke or produce gases more toxic than those produced by wood or paper. The ignition temperature is higher than that of most woods, but acrylic begins to soften above 160°F.

Used for exterior and interior windows, doors, skylights, clerestories and greenhouse glazing, acrylic sheets can be molded into various shapes and contours, as shown in the photos at right. Both single-skin and double-skin versions of the material are available. Single-skin acrylic is clear and comes in sheets or continuous rolls of various thicknesses. Extruded into a hollow-walled sheet material, double-skin acrylic has interior ribs, spaced ⅝ in. apart, running the length of the sheet. It is translucent, but not transparent.

The debate—Builders and designers who have worked with acrylic fall into two camps: they either hate it or love it. Any type of glazing is ultimately compared with glass, and those who like acrylic, whether single or double skin, offer these reasons:

It's versatile. Single-skin acrylic can be cut into a multiplicity of shapes, either for pure design reasons or to meet the demands of an out-of-square solar retrofit. Single-skin acrylic can be cold-formed into curves; both single and double-skin sheets can be heat-formed.

Photos: Valerie Walsh

Acrylic is light, easy to cut, and has more impact resistance than glass, though it scratches easily and has a high rate of expansion. Double-skin acrylic sheets are translucent, and the diffuse light is good for plants. Designer Valerie Walsh thermoforms these sheets into curved roof sections for custom sunspaces.

Acrylics are easy to cut and can be site-fabricated. Lighter in weight and easier to carry, the double-skin sheet is more convenient to install than glass. The flexibility of single-skin sheets varies with their thickness; longer sheets of thinner acrylic require more people to handle them.

Acrylic has high transmissivity, and better impact strength than glass. Double-skin acrylic has an R-value competitive with insulated glass units. It is safe in overhead applications, because it will not shatter. Instead it breaks into large, dull-edged pieces.

On the other hand, builders who prefer glass offer these reasons: Acrylic has a high rate of expansion and contraction, requiring careful attention to keep an installation leak-proof. As it moves, the acrylic sheet makes a noise described as ticking or cracking. And acrylic scratches easily. The extent to which this is considered a problem, or even an annoyance, varies from builder to builder.

Costs—Acrylic used to be much cheaper than glass, but now single-skin acrylic is competitive only when purchased in bulk. Any cost advantage is likely to be lost if you attempt to double-glaze with single-skin acrylic. This is a labor-intensive process, and it's tough to eliminate condensation between the panes. The price of double-skin acrylic is close to that of insulated glass, but the cost is higher if the price of a compression fastening system is figured in. (Some builders expect the price of double-skin acrylic to drop in the future when more companies begin to manufacture it.)

If acrylic is used for a roof in a sunspace or greenhouse, however, its availability in assorted lengths can cut down on labor costs. Designer Larry Lindsey, of the Princeton (N.J.) Energy Group (PEG), points out that long pieces eliminate the need for horizontal mullion breaks, and so can be installed less expensively than several smaller ones. They're also cheaper. An uninterrupted piece of glazing can run the full length of the slope, supported by purlins underneath.

Professional use—Architect David Sellers of Sellers & Co., an architectural firm in Warren, Vt., explains his extensive use of acrylic: "Our whole plastics experiment has been an aesthetic means of expanding the type of architecture

From *Fine Homebuilding* magazine (August 1982) 10:30-33

we do. With acrylic we could push the house beyond what it was already, both the inside and the outside experience of it." In the process, the firm has developed a spectrum of applications for single-skin acrylic (sidebar, below right).

Designer-builder Valerie Walsh, of Solar Horizon, Santa Fe, N. Mex., uses double-skin acrylic for a portion of the roof in the custom-designed greenhouses and sunspaces that are her firm's specialty. She first used single-skin acrylic because it was slick, clean-looking, and didn't degrade in the Southwestern sun. She began to explore unusual shapes, such as a wheel-spoke roof design. Then she turned to using double-skin acrylic. Walsh thermoforms acrylic in her own shop—curved pieces that are as wide as 5½ ft. and typically 6 ft. to 7 ft. long, although she has done 8-footers.

Safety and economics figured prominently in the Princeton Energy Group's decision to use acrylic glazing overhead in their greenhouses and sunspaces.

"The whole issue is a matter of expense," says Larry Lindsey. "In order to have glass products we feel comfortable installing overhead, we have to pay two penalties, one in transmittance and one in bucks. At present, there is no laminated low-iron glass available at a reasonable cost."

For those who have years of experience with acrylic, a willingness to experiment and to learn from mistakes has produced a valuable body of knowledge about working with the material, its design potential and its limits.

Movement—Leaking is a particular concern with acrylic glazing because it moves a lot, expanding and contracting in response to temperature changes. To avoid leaks, design principle number one is to try to eliminate horizontal joints. And wet glazing systems that may do a perfect job of sealing glass joints will not work at all with acrylic. Its movement will pull the caulk right out.

"Acrylic has a tremendous coefficient of expansion—you have to allow maybe an inch over 14 ft. for movement," cautions Chuck Katzenbach, construction manager at PEG, where they have worked with double-skin acrylic for exterior applications and single-skin for interior ones. "No silicones or sealants we know of will stretch that potential full inch of movement." Indeed, one builder tells a story about using butyl tape for bedding: The acrylic moved so much in the heat that the tapes eventually dangled from the rafters like snakes.

Room for expansion must be left on all four sides of an acrylic sheet, because the material will expand and contract in all directions. The amount of movement depends on the length of the sheet and the temperature extremes it will be subject to.

Acrylic glazing can be installed year round, but it is vital to pay attention to the temperature when it is put in place. Katzenbach explains that if it's 30° outside, then you have to remember to allow for expansion to whatever you figure your high temperature will be. If it could go from 30°F to a peak of 120°F in your greenhouse, you have to make provisions for a

Photos: David Sellers

The pliant possibilities of acrylic. **For several years now, the architects at Sellers & Co. have been toying with supple single-skin acrylic to carve shapes and sculpt spaces that abandon the simple linear notion of a house. The concepts that have developed, both successful and unsuccessful, are abundant: sliding doors and windows, cylindrical shower stalls, fixed curved windows, curved windows that spring open at the bottom, bus-style fixed windows, skylights, removable windows held in with shock cords, a continuous window up the front of a house and back down its other side, and an entire roof double-glazed with ⅜-in. acrylic.**

In the late 1960s, curved windows, shapes that would curl in and out from a house, began to fascinate Sellers. First he tried using a heat lamp to form a 12-in. radius curve in a ⅜-in. acrylic sheet. It worked, but the heat produced some distortions. Then he found he could cold-form the sheet into an absolutely clear curved window, just using the building's structure to hold the sheet in the desired shape (photo, top). The first attempt at curving a piece of 3/16-in. acrylic into a shower stall, 8-ft. tall and 6-ft. in diameter, revealed that curving acrylic gave it amazing strength.

From windows, the designers turned to curving continuous sheets of acrylic, some as long as the 45-ft. strips on the sculpture studio at Goddard College, in Plainfield, Vt. The strips stretch from the peak of the roof down to a slow bend at the bottom edge. The limits of curving acrylic became apparent: in a long vertical piece with a curved bottom, the sheet is exposed to incompatible bends, horizontal on the top and then vertical on the bottom, a situation that leads to cracking where the curve begins. In addition, there is the stress of the predominantly vertical movement of such long pieces.

At the Gazley House (photo above), a new detail was tried to support the curve and prevent the cracking. The center bay of the house is glazed with three 23-ft. long strips of single-skin acrylic, fastened with a commercial compression system. Inside and right behind the glazing, a sries of 6-ft. tall wooden ribs, similar in appearance to inverted wishbones, support the curve. The carved wooden ribs double as planters. For an extra measure of support, the designers also installed a horizontal metal crosspiece under the acrylic at the point where the curve begins.

The experimentation with acrylic continues: The firm has begun using a strip heater to bend the edges of single-skin acrylic. Installing adjacent strips would require simply bending up the side edges and then capping the adjacent ones after they were in place. Site-fabricated skylights with wrapped edges would be simple to finish and much less expensive than commercial units. The ideas are just beginning to suggest themselves, and designer Jim Sanford thinks the potential could revolutionize what they do. *—E.H.*

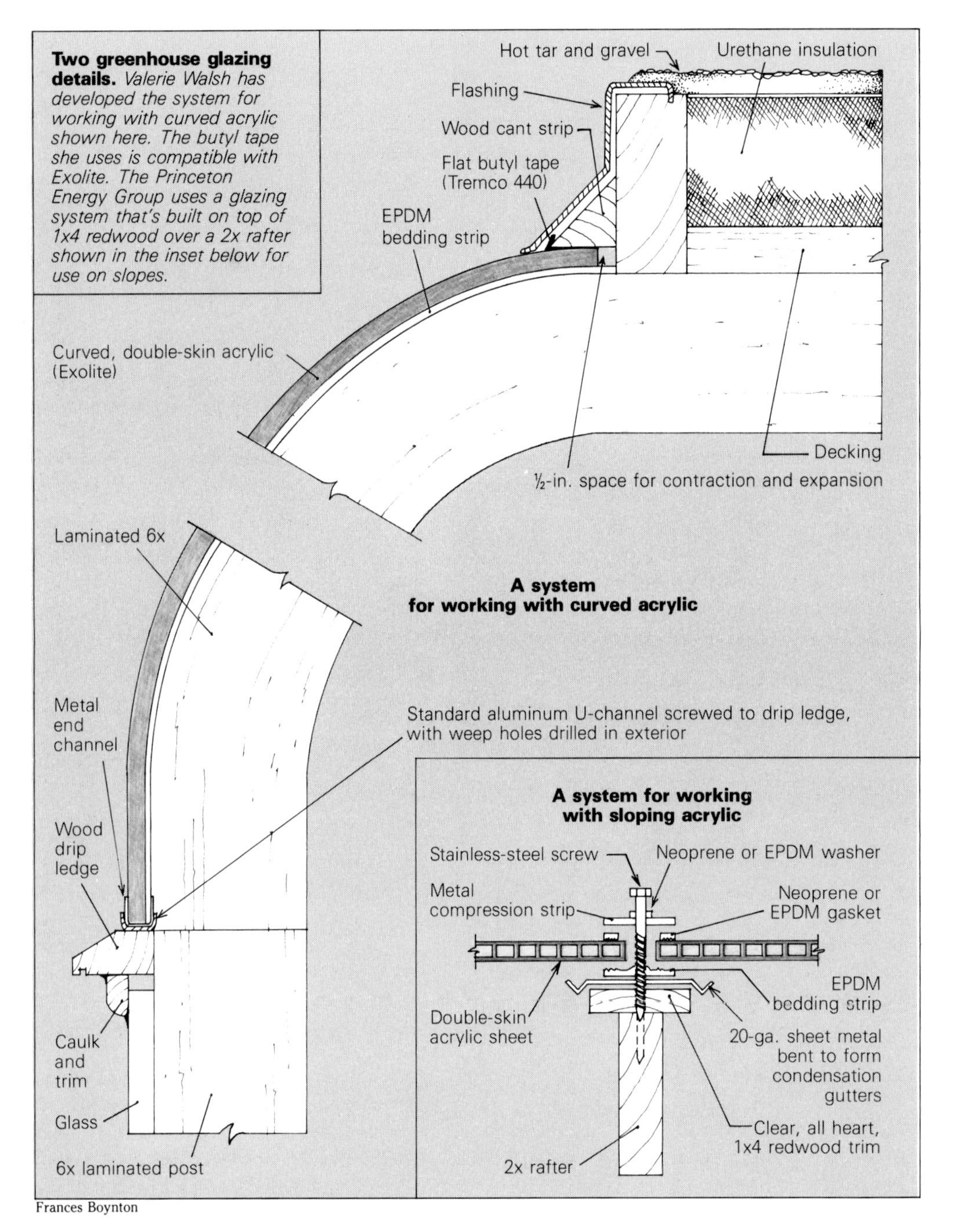

90° change. PEG uses the formula below to calculate the extra space to allow for double-skin expansion:

$$\frac{K}{L} \times \Delta F, \text{ where}$$

K = coefficient of expansion (manufacturer's specs)

L = the length of the glazing (in inches)

ΔF = the difference between the lowest and highest temperatures you expect.

Cutting—To cut a sheet of acrylic, use a fine-tooth carbide-tipped blade set for a shallow cut, and move like a snail. This is important because speed will cause little pressure cracks to appear on the bottom edge. While cutting, make sure that the sheet is firmly supported on both sides of the cut. Sharpness is vital, so use that blade only for working with acrylic. When the acrylic is cut, it heats up and the edges melt, but the wider kerf of a carbide blade will prevent the newly cut edges from melting back together again. After the cut, the edges can be planed, filed or sanded.

As the acrylic is cut, little fuzzy pieces will fly up. Some will reglue themselves to the edges and can be broken off when the cutting is completed. With double-skin acrylic, the fuzz tends to fill the ⅝-in. dia. columns between ribs. Use an air gun to blow it out.

If you drill acrylic, support the sheet fully, and use a very sharp spade bit, ground to a sharper angle than for drilling wood. The sharper angle helps prevent cracking. Drill very slowly, and slow down even more just before the drill breaks through the sheet. Be prepared to break some pieces, no matter how careful you are.

Acrylic sheets come protected with an adhesive masking, which exposure to rain or sunlight makes quite difficult to remove. Leave the protective masking on the acrylic as long as possible, and be prepared for a good zap from static electricity when you pull it off.

Fastening—For years it has been common practice to fasten single-skin acrylic by screwing it down. The designers at Sellers & Co. developed a pressure-plate fastening system to distribute the pressure evenly, and drilled the holes for the bolts or screws an extra ⅛ in. wide to allow for movement. But after ten years or more, the hole has shifted and started pushing against the screw in some installations. Cracks developed where none had existed.

Small cracks in single-skin acrylic can be stopped if the force on them isn't too great. Although his firm now uses installation details that don't involve drilling the sheets, Jim Sanford at Sellers & Co., recommends stopping cracks by drilling a ¼-in. dia. hole at the end of

Acrylic sheets can be cut on site with a sharp, fine-tooth, carbide-tipped blade in a circular saw. Work very slowly to avoid pressure-cracking, then plane, file or sand the new edges smooth. Be sure to support the sheet along both sides of the cut. Photo: Flora LaBriola.

the crack and filling it with silicone. Designer-builder Alex Wade, of Mt. Marion, N.Y., who still uses screws, suggests drilling a tiny hole at the end of the crack, too. He then widens the crack slightly with a knife and fills it with silicone. Finally, he removes the offending screw. Wade suspects that many builders don't take into account the season of the year in which they are working when they drill the holes for screws. When installing acrylic in the extremes of summer or winter, Wade drives the screw either to the inside or the outside of an oversized hole in the acrylic, to allow for subsequent contraction or expansion when the temperatures change. It's important to space the holes evenly (about 2 ft. apart) and to tighten the screws uniformly to distribute the pressure equally. The sheet must be held down firmly, but still be able to move.

To avoid taking a chance with cracking, however, most builders have abandoned screws. Instead they use a compression system of battens that hold the plastic sheet down on a smooth bed of ethylene propylene diene monomer (EPDM). It comes in strips and is supplied by several manufacturers. The acrylic can easily slide across the EPDM as it moves.

Manufacturers recommend a 3x rafter to support the bedding in the compression glazing system. PEG installs an interior condensation gutter on greenhouse rafters that doubles as a smooth, uniform bed for the EPDM gasket in the glazing system. The 20-gauge sheet metal straddles the rafter and is bent into a ⅝-in. lip for the gutter on each side.

PEG has also developed a system for a standard 2x. A clear all-heart redwood 1x4 trim piece is screwed on top of the 2x, widening the bed and providing a smooth surface (inset drawing, facing page). Concentrated stress on the acrylic sheets is as important to avoid with a compression system as it is with screws. If one point is fastened tighter than the others, the acrylic will bow in and leak or crack.

Larry Lindsey recommends aluminum battens on south-facing roofs, because wooden ones will eventually cup upward, creating a leak. Aluminum battens can be purchased with various finishes, or they can be capped with a strip of redwood.

Double-skin acrylic needs to be supported at its base or it will bow instead of moving within the compression glazing system.

PEG lets the sheets hang over the roof's edges as a shingle would, sealing it underneath. For the bottom edge on installations with curved roofs, Valerie Walsh has developed a system with no damming problems. She slides on aluminum terminal section (ATS) from CYRO (697 Route 46, P.O. Box 1779, Clifton, N.J. 07015) on the bottom of the double-skin acrylic, then snugs the acrylic into a larger aluminum U-channel that is in turn screwed into the wood beam. Walsh then caulks the inside and drills weep holes through the outside of the U-channel (large drawing, facing page).

In one of his designs, David Sellers decided to glaze the south-facing roof area with long strips of single-skin acrylic. To avoid leaking, he encouraged the tendency of the ¼-in. sheets to sag slightly. Small blocks under the edges of the sheets accentuate the dip. Melting snow or rain flows to the center of each panel and then drains off the roof. On the bottom edge, an angle keeps the acrylic from slipping, as shown in the photo above.

Single-skin acrylic is flexible, and some designers take advantage of this to encourage drainage. Extra blocking along the sheets encourages a sag that carries water away. Photo: David Sellers.

Glazing materials—Acrylic is fussy stuff. Chuck Katzenbach reels off a list of materials to avoid using with this plastic. Vinyl leaches into acrylic and weakens its edges. Some butyls have plasticizers that may also leach into acrylic. In these cases, either the acrylic will eventually fail or the butyl will become very hard. The plasticizer in most neoprenes is not compatible, so check with the manufacturer.

Compatible glazing materials are few: EPDM heads the list. Silicone caulk is okay, but sooner or later the acrylic's movement will pull it loose. If it's installed on a cold day, the silicone may pull out on the first really warm one. There are many urethane foams you can use, but you would be well advised to consult the manufacturer directly about compatibility.

Support—The double-skin acrylic can bow in over the length of the roof, and the sheet could conceivably pull out of the glazing system under a very heavy snow load, according to Larry Lindsey. So as a cautionary measure, PEG figures on a 30-lb. snow load and installs purlins 4 ft. o.c., about ⅝ in. below the sheet.

Condensation—Acrylic transpires water vapor, so a double-skin unit typically will have some cloudy vapor inside. Double-skin acrylic should be installed with its ribs running down the slope so that any condensation inside the channels will collect at the bottom edge of the sheet. This edge needs to be vented. Double-skin sheets arrive with rubber packing material in both ends of the channels, to keep them free of debris. PEG's construction crew just leaves it there and perforates it with a scratch awl to allow air movement.

Cleaning—Never use abrasives, ammonia-base glass cleaners or paint thinner on acrylic glazing. Mild detergent, rubbing alcohol, turpentine and wax-base cleaner-polishers designed especially for plastic are safe when applied with a soft cloth.

Cleaning brings up the controversial issue of scratching. "The scratching drives me crazy, though other builders don't seem to care as much," says Valerie Walsh. Whether she is storing the sheets she has heat-formed into curves or transporting them to the building site, she keeps thin sheets of foam padding wrapped around every piece.

"With double-skin acrylic, people's tendency is not to expect to be able to see through it. They're not looking through it as they would through a window, and so they're not seeing the small imperfections in the surface itself," argues Chuck Katzenbach.

Even though single-skin acrylic is transparent, many builders who work with it say that scratching just isn't a significant problem, particularly if the glazing is kept clean. The only serious scratching problem is likely to be the work of a dog. Most scratches can be easily removed with a Simonize paste-wax buffing. And now some single-skin acrylics are available with a polysilicate coating that makes them abrasion-resistant and also improves their chemical resistance. □

Elizabeth Holland, of West Shokan, N.Y., is a contributing writer to this magazine.

Sources of supply

Acrylite (single), **Exolite** (double): CYRO, 697 Route 46, P.O. Box 1779, Clifton, N.J. 07015.

Lucite (single): DuPont Co., Lucite Sheet Products Group, Wilmington, Del. 19898.

Plexiglas (single): Rohm and Haas, Independence Mall West, Philadelphia, Pa. 19105.

Acrivue A (single, with abrasion-resistant coating): Swedlow Inc., 12122 Western Ave., Garden Grove, Calif. 92645.

Attic Venting

Installing vents can keep home heating and cooling costs from going through the roof

by William R. Wheeler

Most of today's "tight" houses, built or remodeled to conserve as much energy as possible, also need the benefit of proper attic ventilation. Provisions for good ventilation are neither difficult nor expensive, and can save the home owner energy dollars and prevent damage to insulation and structural members.

Proper ventilation combines natural wind and thermal forces to remove heat from the attic in the summer and moisture in the winter. In an unventilated attic, summer sunshine on the roof heats the shingles and roof sheath. These materials can get as hot as 165°F to 170°F. Insulation in the attic floor joists is slowly heated and may reach 110°F to 120°F. Convection currents in the attic move air from warm surfaces, and in a relatively short period of time, attic air is up to 130°F. Thus the ceiling warms and radiates heat to the living space below.

During the winter, water vapor makes its way into the attic by vapor pressure (humid air diffusing into areas of lower humidity) and condenses on the roof sheathing as frost. The thaw/freeze cycle and continual moisture build-up causes the roof framing and sheathing to rot. Water can condense on the rafters, drip onto the insulation and then damage the ceiling below. Outside the house, peeling or flaked-off paint on the soffit is a sign of moisture build-up behind the soffit boards. The average family of four puts 18 lb. of moisture into the air daily; washing machines and clothes driers, stoves, humidifiers and moist basements or crawl spaces add to the substantial amount of water vapor that can be trapped in a poorly ventilated attic.

Vents create air movement—When air is removed from an attic, it has to be replaced. To create the air movement that is the basis of good ventilation, most houses with gabled roofs need soffit vents under the eaves and a ridge vent at the peak. During warm weather, warm air rises naturally to the roof peak and escapes through the ridge vent. The exhaust action draws fresh air in through the soffit vents (drawing, below left). This air picks up heat from the roof as it rises under the roof sheath and passes out the ridge vent.

In winter, a properly ventilated attic will be only 3°F to 10°F warmer than the outside air, so the moist house air doesn't condense inside the attic, as it would in a warmer, poorly ventilated space. The warmth held inside an unvented attic can melt snow on the upper section of the roof, and this meltwater will re-freeze on the cooler lower roof above the soffit. A cool attic is necessary to keep snow on the roof and prevent ice build-up. With the recommended R-38 attic insulation in the ceiling, heat loss from the living space below will be kept to a minimum. In fact, tests have shown that increased attic ventilation does not increase winter heating costs.

There are alternatives to the ridge and soffit vent configuration for venting your attic. For example, roof and gable-end louvers can be used independently or in combination with soffit vents. And if the roof of your house has no peak, then soffit vents and roof louver vents are what you need. The important thing is to establish a flow pattern in the attic that will eliminate dead spots of stagnant air.

How large a vent do you need?—The standards for attic ventilation in the chart on the facing page are based on a ventilation rate of 1.5 cfm (cubic feet per minute) per sq. ft. of attic floor area. A vent system that lets air circulate at this rate should keep heating and cooling costs

Illustrations: Barbara Smolover

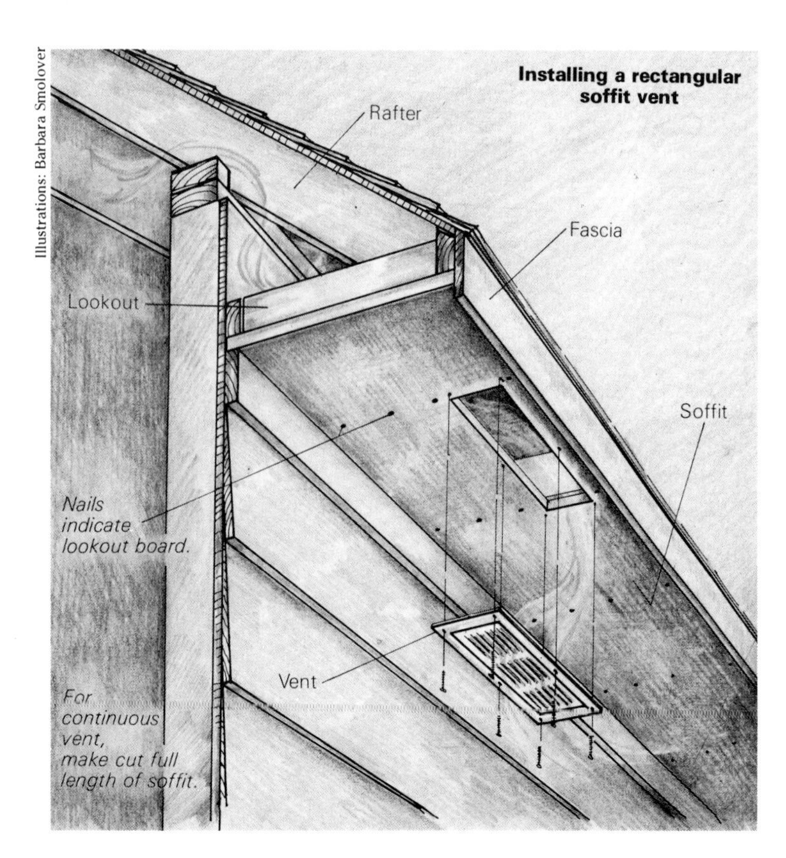

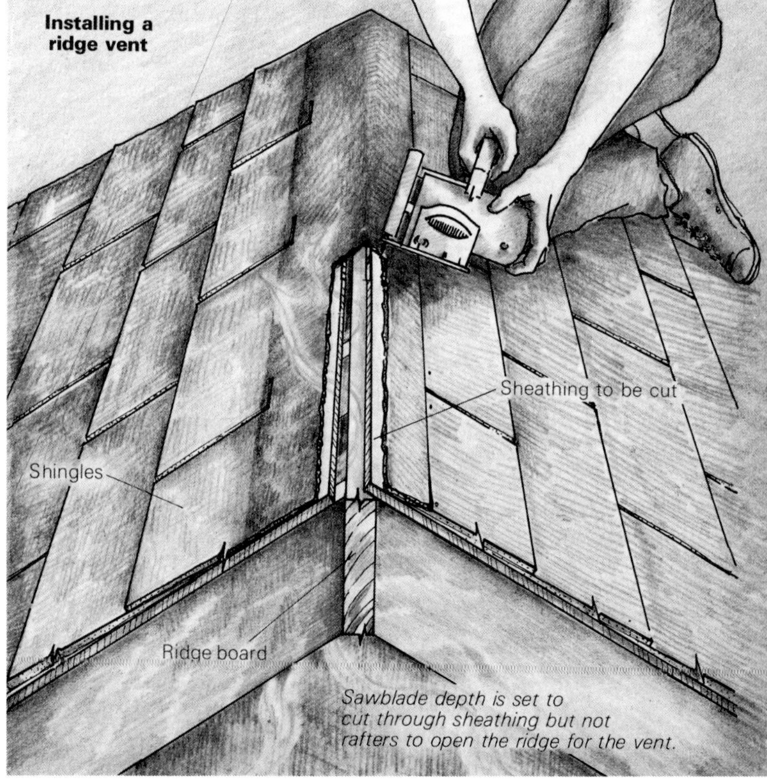

From *Fine Homebuilding* magazine (February 1982) 7:52-53

down, and minimize the possibility of water damage due to condensation in the attic.

To determine how large your vents should be, first calculate the square footage of your attic floor and locate this figure on the baseline of the chart. Then follow a vertical line upward to the point where it intersects the line of the venting system you'll be using. Read horizontally to the left for the net free ventilation area required.

Net free ventilation area (NFVA) is a measure (in square inches) of how much air a given vent will let through. This figure is usually printed on the vent. An average-sized ridge vent has an NFVA of 18 sq. in. per lineal foot. If you plan to install screening over soffit openings instead of using manufactured vents, add 25% to the NFVA figure to determine the dimensions of your soffit openings. Remember that the NFVA should be divided equally between intake and exhaust vents. In a conventional gable-roofed house, for example, half the total NFVA would be allocated to the ridge vent, with soffit vents each having one-quarter of the total figure.

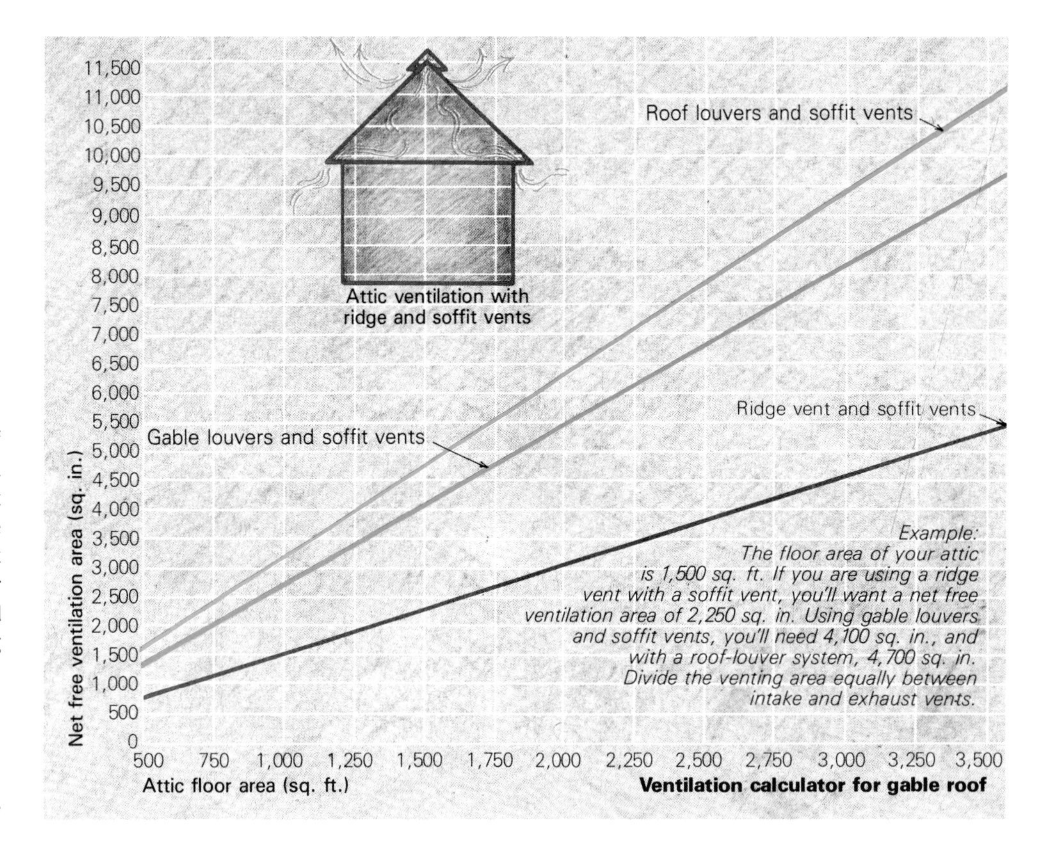

Ventilation calculator for gable roof

Installing a soffit vent—You can either cut the openings and cover them with screen, or fit factory-made units into the soffit. In both cases you'll have to saw out openings in the soffit board, using either a keyhole saw, a saber saw or a circular saw. Watch out for lookout boards, since cutting into them will weaken the soffit. Lookout boards are usually located on centers identical to those of the rafters; the best way to find them is to look for several nails in a line across the width of the soffit. If you're using a circular saw, adjust the blade depth to the thickness of the soffit. Be sure to remove all nails that will be in the path of the blade.

Continuous soffit vents run the entire length of the soffit. Match the width of the slot to the fit required by the continuous aluminum vent. Leave at least 2 in. between the outside edge of the soffit and the outermost cut-line. One good way to mark cut-lines is to snap parallel chalklines from one end of the soffit to the other. If you are planning to cover the slot with hardware cloth or some other kind of screening, 2 in. to 3 in. is a good width.

As an alternative to a continuous soffit vent, there are rectangular vents that you can install between lookout boards. You'll need to install a series of these for adequate ventilation. The easiest way to mark cut-out locations is to use a cardboard template that corresponds to the fit of the vent. Drill holes in the corners of each rectangle and saw out the sides with a saber saw or a keyhole saw. Then lift the vent into position and nail or screw it in place.

Some soffit vents have individual louvers that slant uniformly toward one side of the unit. If this is the type you're working with, be sure that they slant away from the house, toward the outside of the soffit. This directs incoming air up against the roof sheath, where the circulation will do the most good.

If you choose to screen your soffit openings, the best time to do so is during construction. This way, you can tack the screen to the inside face of the board before the soffit is nailed in place. Otherwise you'll have to fasten the screen over the soffit slots on the outside face of the board. A screen with less than ⅛-in. openings is best. Be sure to tack or staple the screen securely; otherwise birds and bees find their way into soffit nests. Trimming the vents with molding will keep them out. It also looks better.

Installing a ridge vent—A ridge vent is designed to fit over a long, narrow opening at the peak of the roof. For new house construction, you can leave this space open when sheathing and shingling the roof. The width of the slot will depend on the dimensions of the vent. To retrofit a ridge vent in a gabled house you have to remove the cap shingles along the ridge (drawing, left). Then snap chalklines on opposite sides of the roof, running parallel to the ridge and 1½ in. to 2 in. down from the peak.

Use a utility knife to cut through the shingles and roofing felt along these two lines and strip away the cut-out material to expose the sheathing. Cut away the narrow bands of exposed sheathing on either side of the roof peak with your circular saw. Take care to pull all nails in the path of the saw, and set your depth adjustment so that you'll cut only through the sheathing and not into the rafters.

A pry bar is handy for removing the strips of sheathing once you've ripped through them with your saw. The slots you've opened up between rafters and on either side of the ridgeboard will serve as air passageways up through the ridge vent. You'll have to cover the upper half of the shingles exposed when the cap course was removed. Use matching shingles or other good-quality roofing materials to protect the peak from leaks during storms. Cut them narrow enough so that the top edge comes up to the slot, leaving the bottom edge exposed.

Ridge vents come in sections that must be fit together to make a continuous run across the roof peak, using connector caps to cover the joints. Rather than nail each section to the roof individually, it's better to put the entire assembly together and align it on the ridge before nailing. Otherwise you may get an uneven, staggered appearance. Properly installed and painted to match the roof color, ridge vents can be unobtrusive as well as functional. □

Bill Wheeler, of Princeville, Ill., is technical manager of HC Products Company, manufacturers of attic-venting equipment.

Heat-Recovery Ventilators

These vital appliances for superinsulated houses exchange stale air for fresh

by John R. Hughes

The trend toward tightly sealed, superinsulated houses designed to cut heating bills has brought about the development of air-to-air heat exchangers—now commonly referred to as heat-recovery ventilators, or HRVs. Large industrial HRVs have been in use for many years, so you might assume that downsizing them to suit a house should have been a fairly painless task. This unfortunately has not been the case. Of all the components that are necessary in today's energy-efficient houses, HRVs seem to be getting the worst reputation. A few examples will illustrate the problem.

After the first year of the Residential Standards Demonstration Program in Washington, Oregon, Idaho and western Montana, many contractors singled out HRVs as the most troublesome item they had to deal with. Problems included cost, inadequate information on installation and balancing, poor technical support from manufacturers and the constant need to defrost.

Similarly, in a 1982 home-owner survey done in Canada (released in 1984), almost all of the 147 participating home owners reported one or more problems with their HRVs. These ranged from noise and substandard installation to malfunctioning controls.

As if to add insult to injury, the Ontario Research Foundation tested the performance of 12 models of heat exchangers in 1983-84. They checked airflows, cross-contamination, heat transfer in cooling conditions, heat transfer in non-frosting heating conditions and perfor-

Ductwork for heat-recovery ventilation with baseboard heating system

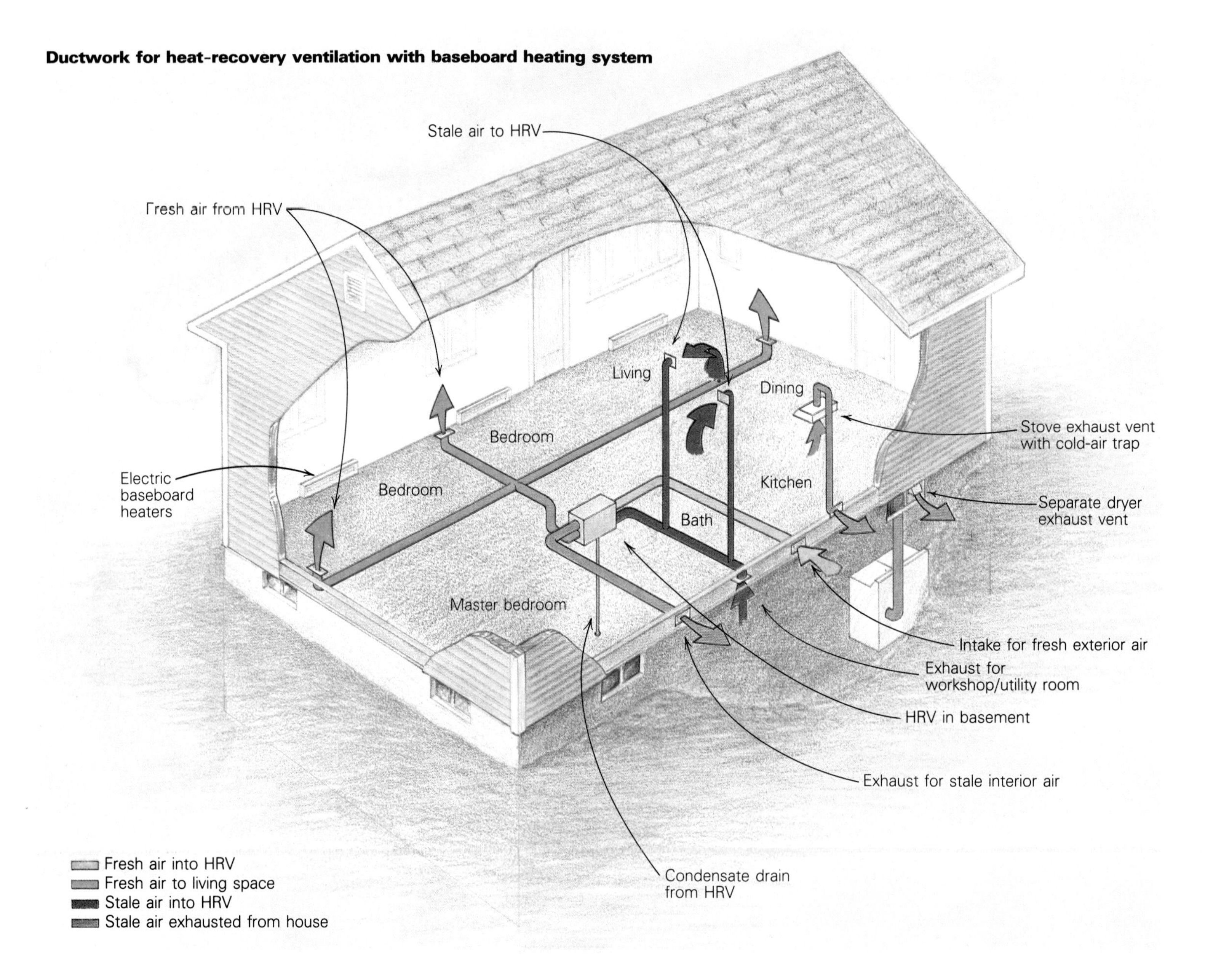

From *Fine Homebuilding* magazine (August 1986) 34:30-34

mance in cold-weather conditions. When the results were first published in early 1985, it appeared that none of the HRVs performed very well, and that some had failed certain parts of a performance test.

Why all these problems with HRVs? To answer that question, I'll first explain what HRVs are supposed to do, and how they do it.

The need for ventilation—Although they are commonly called heat exchangers, the main function of HRVs is to get rid of the odors, excess humidity and indoor air pollutants that can easily accumulate in a tightly sealed house. By tightly sealed, I mean a house whose vapor barrier is so continuous that the natural air-change rate (how much air enters and leaves the house through cracks and openings) is less than ½ air change per hour. Leaky houses (with air-change rates above 1 per hour) are less energy efficient, but don't usually present a great potential for indoor air-quality problems.

HRVs are designed to exhaust stale air from parts of a house where moisture, odor or pollutants are likely to accumulate. Fresh air is drawn into the house through the HRV. Stale air also passes through the HRV on its way out of the house. HRVs are designed so that the outgoing air can pass some of its heat (during winter operation) over to the incoming air.

Before going any further, we should lay to rest a few oft-heard criticisms of the tight-house concept. First, why go to all the trouble and expense of tightly sealing up a house and preventing air movement, and then put in an expensive machine to overcome the ill effects arising from the airtightness? One might as well ask why put powerful engines in cars, and then add complex power-braking systems to overcome the speed developed by the engine. HRVs, like power brakes, give us more control and a greater degree of safety.

Second, why install a mechanical ventilation system when opening a window will provide fresh air? The problem with this is that the open window only provides fresh air to a very small, localized area of the house. An HRV exchanges the air in all parts of the house, even those seldom used rooms where the door is normally kept closed. And, of course, an open window lets in air at outside temperatures. The HRV tempers incoming air, an important feature in the winter. Residential-sized HRVs were developed in Saskatchewan in the late 1970s. Winters there are long and cold. For at least a few weeks every year, the temperature hovers well below −35°F. In the northern part of the province, −50°F is not unusual. With temperatures like that, one doesn't want a lot of super-cold air blowing in through the walls to replace stale air being kicked out by the kitchen or bathroom exhaust fans.

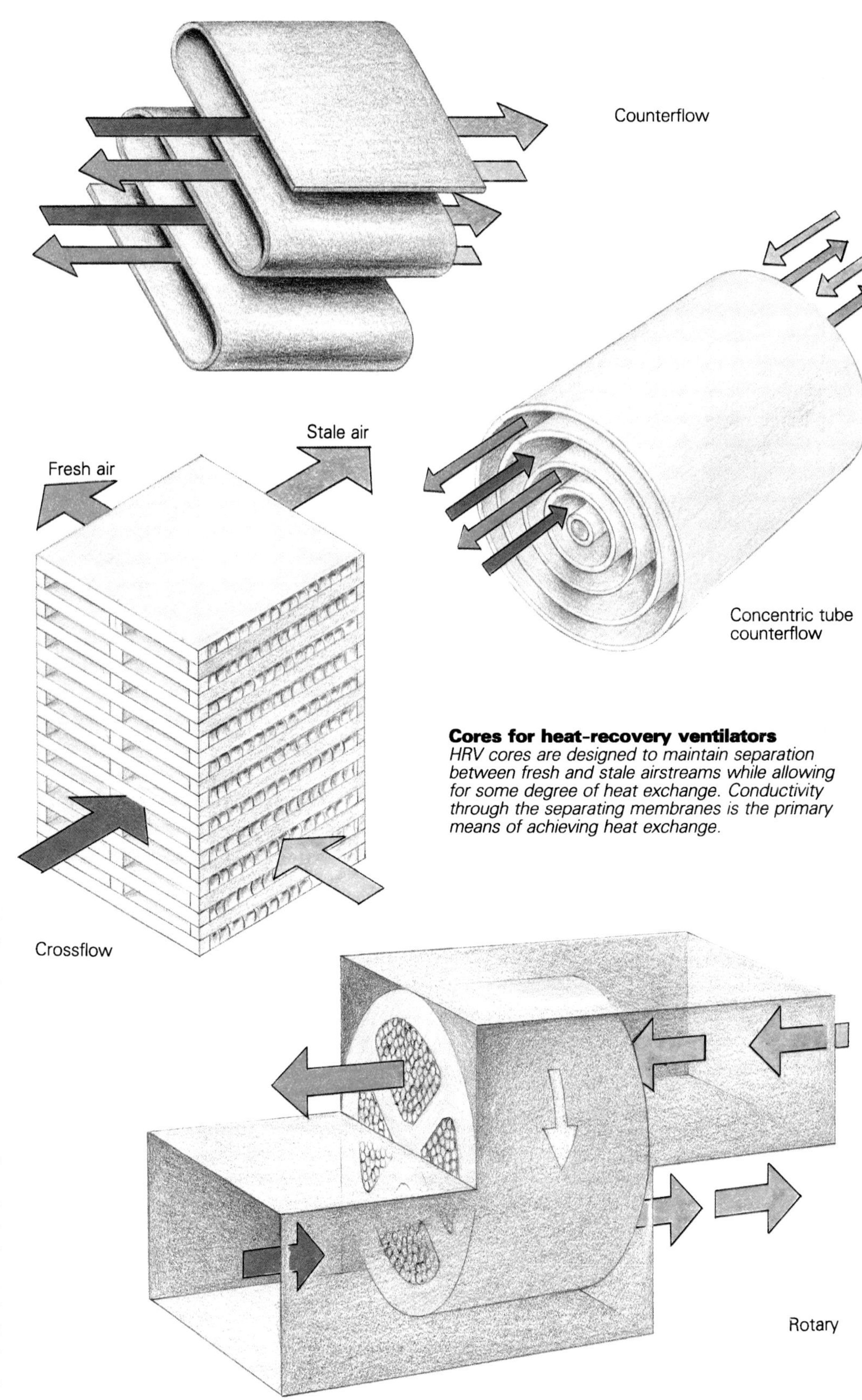

Cores for heat-recovery ventilators
HRV cores are designed to maintain separation between fresh and stale airstreams while allowing for some degree of heat exchange. Conductivity through the separating membranes is the primary means of achieving heat exchange.

How HRVs work—An HRV draws stale air from several different locations in the house—usually kitchens, bathrooms, spa or hot-tub areas and hobby areas—and exhausts it through a single outlet in an exterior wall. The single outlet minimizes penetrations in exterior walls. To replace this stale air, the unit draws fresh air from outside (through a single inlet) and distributes it throughout the house (drawing, facing page). Kitchen vent fans operate independently of the HRV to prevent high concentrations of smoke and oil from accumulating in the core of the HRV and reducing its efficiency.

In a typical HRV, exhaust and intake airstreams are ducted through a two-directional, multi-channeled plenum. They are kept apart by thin membranes of plastic, metal or treated paper. The membranes prevent mixing of the two airstreams to prevent cross-contamination or cross-leakage, yet they allow heat from the outgoing warm air to transfer by conduction to the incoming cold air. The plenum is arranged so that next to each incoming airstream is an exhaust airstream. This warm-cold-warm-cold arrangement ensures maximum heat conduction, as shown in the drawing above. The closer the membranes are to each other—typically ⅛ in. to ¼ in.—the better the heat transfer.

The heat recovery, though, is a mixed blessing. As the warm, moist, stale air is cooled on its way through the heat exchanger, it loses its ca-

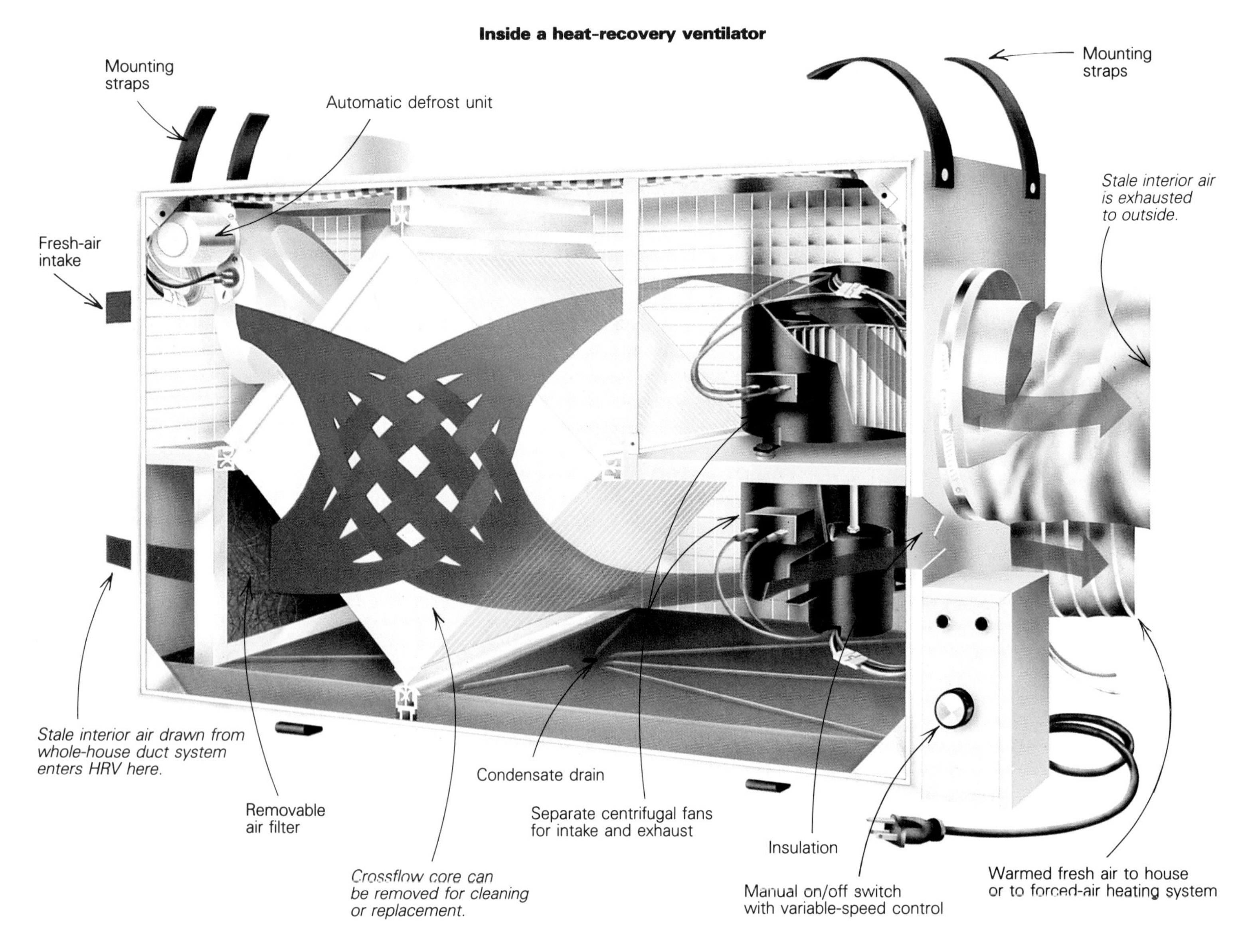

pacity to hold moisture. Condensation within the heat-exchange core requires a drain pan and hose on the warm side. And if the HRV is used in a climate with both cooling and heating seasons, two drains may be necessary.

In a very cold climate, however, condensate can freeze in the HRV core if the outgoing air is cooled below the freezing point. Since the build-up of frost in the air channels reduces the air flow, it's necessary to build in a defrost system. When HRVs were first developed, most were defrosted by shutting off the intake-air fan (there are two fans, one for each airstream) and simply letting the relatively warm outgoing air melt the frost. However, this was found to induce a slight negative pressure in the house, since more air was leaving than was being taken in.

Slight negative pressure in an ordinary house is no problem—there are enough breaks in the air barrier to allow air to move in and re-establish equilibrium. But in a tightly sealed house, there may not be enough openings for this to occur. So the air will try to get in through openings that weren't intended to let it into the house. If there is a fuel-fired heating system, or a fireplace or woodstove, for example, cold outside air may be drawn down the chimney. Back-drafting—air moving down the chimney—will push smoke and carbon monoxide into the house, creating a serious health hazard. With both HRV fans running, pressure in the house remains balanced.

Several solutions to the defrosting problem have been suggested. One is to put the HRV in a sealed room along with the furnace, and vent the room to the atmosphere. This idea originated in Saskatchewan, where the provincial building code requires atmospheric venting for fuel-fired appliances such as furnaces. In an electrically heated house with a woodstove, however, the problem remained the same. Some manufacturers have incorporated an electric heater in the HRV core. A sensor detects frosting, and activates the heater. This is a wise choice except in climates where long, cold winters are the rule, or where electric rates are high. A third solution is to close off the intake air line temporarily and create an opening in the intake line within the house itself. Thus warm air is pumped through the core in both directions, cutting defrost and negative-pressurization time to a few minutes.

Core considerations—Several factors influence core design. One, of course, is the spacing between the membranes. An HRV with twenty ⅛-in. thick airstreams will have better heat-transfer characteristics than one with ten ¼-in. airstreams. Equally important is air speed—the more slowly the two airstreams move past each other, the better the heat transfer. Decreasing air speed too much, though, may reduce the air-change rate below the required level. And of course, the larger the area of the core, the better the heat transfer, assuming all other factors remain the same. But size can be a problem for installers trying to fit a large HRV in a small room. The core material also affects heat transfer—metal conducts heat better than plastic, but costs more. Decreasing membrane spacing, slowing air speed, using a metal core, and increasing the core area all raise the efficiency, and the cost, of an HRV.

Having taken care of the considerations of membrane spacing, air speed and core size and material, there arises the question of core configuration—the path the air will follow as it goes through the HRV. There are several configurations possible (drawings, previous page), each with its own advantages and disadvantages.

Counterflow. In a counterflow core the two airstreams move parallel to each other, but in opposite directions. In theory, 100% of the heat from the exhaust air can be transferred to the replacement air in this type of core. In practice, of course, this is impractical, as the core size would have to be very large, and the air speed very slow. Counterflow cores are also difficult to design and construct without a lot of cross-contamination problems.

One HRV manufacturer, Blackhawk Industries,

Drawing: Courtesy Conservation Energy Systems

Inc. (see the address information on the next page), has developed an interesting variation on the normal flat-plate counterflow core. Their heat-exchanger core relies on concentric tubes, with the air moving in opposite directions from one ring to the next.

Crossflow. This is a very common type of core arrangement. The two airstreams move past each other at right angles, often looping back through a second crossflow core. Single-core HRVs (like the one in the drawing on the facing page) generally aren't as large, expensive or efficient as double-core units. Generally, crossflow cores are less expensive to build than counterflow cores.

Rotary. This is the most compact type of core on the market, and is used by Nutone, among other manufacturers. A rotary core consists of an air-permeable wheel that rotates over the intake and exhaust ports. Stale outgoing air warms the air passages in the wheel, which then rotates to cover the intake port. The cold fresh air is warmed slightly as it passes through this section of the wheel, which continues to rotate back toward the exhaust port.

Capillary. This, like the rotary core, is a regenerator with a spinning action. Two airstreams are forced through a rapidly spinning foam "donut." The air enters through the "hole" and is forced out through the edges, losing heat to or picking it up from the foam. Cross-contamination can be a problem with this system—in one unit tested by the Ontario Research Foundation, 27% of the exhaust air found its way back into the incoming airstream.

Heat pipe. A heat pipe is a metal tube partially filled with a refrigerant (like Freon) that has a very low boiling point. The tubes are usually installed at a slight slope, with the low end in the outgoing airstream, and the high end in the incoming airstream. Heat from the exhaust air boils the Freon, which rises in the tube, loses its heat to the cold incoming air, condenses, and runs back to the exhaust end. There are only a few heat-pipe type HRVs on the market (Q-dot Corp. and Future Energy Products, Ltd., for example), and they tend to be more expensive than plastic-core models. Preliminary research seems to indicate, however, that cross-leakage can be eliminated completely when heat-pipe cores are used in an HRV.

Fans and controls—The type of fan used with a heat-recovery ventilator has a great influence on its performance and cost. The two most common fan types used in HRVs are axial and centrifugal fans.

An axial fan, more commonly called a propeller fan, is installed directly inside the exhaust or intake air duct. They are less expensive than centrifugal or "squirrel-cage" fans, but have to work harder—or be larger—to move the same amount of air.

Vibration from fans has been a major problem with many HRVs. One solution is to isolate the fan from the rest of the ductwork by installing a short length of flexible duct on each side of the fan. Although the flexduct absorbs any vibration, the roughness of its interior surface causes a noticeable pressure drop—a serious consideration where long duct runs are needed. A second solution is to mount the fans directly on—or in—the exchanger case and isolate the case from the house framing with rubber straps or gaskets. If all else fails, mufflers are available for the intake air line.

Instead of being located in the basement, this HRV is mounted on the ceiling of a garage apartment in a superinsulated saltbox.

Most HRVs come with controls either built into the unit or supplied for installation in other parts of the house. The most common types of controls are simple on/off switches, variable-speed controls, timers and dehumidistats. If these are built in or attached to the exchanger, adjusting them can be a bit more trouble than if they are mounted in the living area of the home. However, once a satisfactory control strategy has been worked out, unit-mounted controls are less likely to be tampered with.

The simplest control is the on/off switch. It is usually mounted in a central location within the house and is used to run the exchanger only when the home owner turns it on. The shortcoming of this type of control is that the home owner may not know when the unit should be running. This can be partially overcome by wiring the on/off switch in line with normal switches and controls within the house, such as the light switch in the bathroom or utility room. Then, as long as the light is on, the exchanger runs, exhausting excess humidity and odors. Adding a timer to switches in high-humidity areas is a good idea, as not all the humidity is likely to be exhausted while the light is on.

Variable-speed controls allow the home owner to run the HRV at low speed most of the time, providing a minimal air-change rate. A higher rate of air exchange can be dialed as pollutant levels rise, as they would during a party when people are smoking. You can also add a dehumidistat. This device senses the relative-humidity level of the house air and kicks the HRV into high speed when the humidity exceeds the level set by the home owner (usually 40% to 50% relative humidity).

System design—As with a plumbing or heating system, an HRV system should be designed for each specific house and its occupants. What follows are some basic rules of thumb.

First, determine the air-change capacity that your HRV should have, given the tightness of your house. To do this, calculate the volume of the entire house (V), including the basement if it is likely to be used as living area.

Determine the total (natural plus HRV induced) air-change rate desired (T). Though no absolute air-change per hour (ACH) rate has been established for superinsulated houses, ½ ACH is generally considered to be acceptable.

Next, have the house tested for its natural air-leakage rate (L). If the house is still under construction, you'll have to estimate L, based on the house design and the tightness of similar houses built by the same contractor.

Subtract the natural air-leakage rate from your ACH target $(T - L)$. Multiplying the house volume (V) times this figure tells you how much air the HRV must be capable of moving per hour. Divide by 60 to get the cubic-foot-per-minute rate. The formula looks like this: $V(T - L)/60$.

Typically, the HRV in a 1,200-sq. ft. house with a full basement should be capable of moving about 125 cfm, assuming 0.1 ACH natural leakage, and 0.4 ACH induced air change. Keep in mind that this is the amount of air that must be moved through a fairly convoluted set of ducts. The rated capacity of the HRV should be larger to make up for the inevitable pressure drop due to ductwork.

Pressure drop can be significant in calculating HRV capacity. Small-diameter ducts and runs that are long or twisty reduce the amount of air that actually gets through a duct system. A single 90° elbow provides the same resistance to air movement as 10 ft. of straight galvanized metal duct. Most HRV manufacturers have guidelines for determining static-pressure drop. There's also some good system-design and layout information in *The Complete Heat Exchanger Book* ($22.90 postpaid, Northern Scientific, Inc., P.O. Box D, Minot, N. Dak. 58702).

To minimize duct runs and pressure drops, the HRV should be situated in a fairly central area of the house. It has to be accessible for maintenance. The most common locations are in a utility room or in the basement near the furnace. A floor drain or DWV connection will be necessary to dispose of condensate. Since all HRVs use electricity, make sure that an outlet is nearby. Most HRVs are not designed to work in an unheated space, such as an attic. Never locate an HRV in a garage area, or near other sources of fumes or pollution.

Although the ideal situation is to have a stale-air pickup or a fresh-air outlet in each room or area of the house, in practice this is usually only approximated. An open-plan house needs fewer pickups and outlets than a similar house that is walled off into smaller rooms, although the total air flow is the same. Minimizing the number of pickups and outlets simplifies the duct design and keeps pressure drops to a minimum, but it can lead to poor air circulation in some rooms.

The bathroom pickup replaces the conventional exhaust fan. It should be located on an interi-

Photo: Howard Faulkner

Air-to-air heat-exchanger manufacturers

ACS Hoval
935 N. Lively Blvd.
Wood Dale, Ill. 60191

The Air Changer Div.
Nortron Industries Ltd.
1140 Tristar Dr.
Mississauga, Ont., Canada L5T 1H9

AirXChange, Inc.
30 Pond Park Rd.
Hingham, Mass. 02043

American Aldes Ventilation Corp.
4539 Northgate Court
Sarasota, Fla. 33580

Aston Industries, Inc.
P.O. Box 220
St.-Leonard d'Aston
Quebec, Que., Canada J0C 1M0

Berner International Corp.
P.O. Box 5205
New Castle, Pa. 16105

Blackhawk Industries, Inc.
607 Park St.
Regina, Sask., Canada S4N 5N1

BossAire
1321 Tyler St. N.E.
Minneapolis, Minn. 55413

Can Ray Inc.
255 Restigouche Rd.
Oromocto, N. B., Canada E2V 2H1

Conservation Energy Systems
(vanEE brand HRVs)
(Canadian) 3310 Millar Ave.
Saskatoon, Sask., S7K 7G9
(U.S.) Box 10416
Minneapolis, Minn. 55440

Des Champs Labs, Inc.
Box 440
East Hanover, N. J. 07936

Enermatrix, Inc.
P.O. Box 466
Fargo, N. Dak. 58107

Ener-Quip Inc.
99 E. Kansas St.
Hackensack, N. J. 07601

Engineering Development Inc.
4750 Chromium Dr.
Colorado Springs, Colo. 80918

Environment Air Ltd.
P.O. Box 1128
Moncton, N. B., Canada E1C 8P6

Flakt, Inc.
Products Division
500 Shepherd St.
Winston-Salem, N. C. 27102

Future Energy Products Ltd.
184 Rocky Lake Rd.
Bedford, N. S., Canada B4A 2T6

Mountain Energy & Resources, Inc.
15800 West Sixth Ave.
Golden, Colo. 80401

NewAire
7009 Raywood Rd.
Madison, Wis. 53713

Nutech Energy Systems, Inc.
97 Thames Rd. East
Box 640
Exeter, Ont., Canada N0M 1S0

Nutone Inc. Scovill
Madison and Red Bank Rds.
Cincinnati, Ohio 45227

P. M. Wright Ltd.
1300 Jules-Poitras
Montreal, Que., Canada H4N 1X8

Q-Dot Corp.
701 North First St.
Garland, Tex. 75040

Standex Energy Systems
P.O. Box 1168
Detroit Lakes, Minn. 56501

Star Heat Exchanger Corp.
B-109 1772 Broadway St.
Port Coquitlam, B. C.
Canada V3C 2M8

Xetex, Inc.
3530 E. 28th St.
Minneapolis, Minn. 55406

or wall to avoid penetrating the vapor barrier. In some cases the stud cavity can be used as a duct, with a metal boot at floor level connecting it to the exhaust line. It's advisable to line the cavity with 6-mil poly to prevent the drywall from absorbing moisture from the exhaust air.

Kitchen pickups should be kept well away from stoves to minimize the amount of grease being ducted into the HRV. The regular kitchen exhaust should be vented outside well away from the HRV fresh-air inlet.

Clothes dryers should not under any circumstances be ducted into HRVs. The high air temperatures could cause melting of some plastic cores, and the terrific amounts of water vapor will cause extreme freeze-up problems in very cold climates. Lint also causes problems in an HRV core.

Filters are a necessity, both on the intake and exhaust air lines. On the exhaust side, they keep dust from clogging the HRV core. On the intake side, they keep dirt, insects and organic matter out of the house. Most filters supplied by HRV manufacturers are simple fiber-mesh units, similar to those sold for furnaces. High-efficiency filters for taking pollen, smoke and fungi out of the air are also available.

Stale-air exhaust and fresh-air intakes should be located at least 2 ft. above the ground (higher in areas with lots of snow) and at least 6 ft. apart. All ducts on the "cold" side of the HRV should be insulated to prevent moisture from condensing on them.

Installation and balancing—Having an HRV system properly sized and laid out is wasted effort if the installer doesn't follow the plans. If the plans call for a straight pipe and one 90° corner, make sure this is actually what's built. The installer may find it easier to put in a couple of jogs and run the line where he has more headroom to work in—but the finished system won't work the way it's supposed to. Choosing an installer is just like picking any other subcontractor. Look for one who has a good reputation. Ask for references and, if possible, go see some installations he has done.

In Canada, any installer who is working on an R-2000 house must have taken a two-day training course conducted by the Heating, Refrigeration and Air Conditioning Institute (5461 Dundaf St. West, Suite 226, Islington, Ont. M9D 6E3). The HRV installation itself must conform to Canadian Standards Association Installation Guideline C444.

Most of the exhaust air is drawn from kitchens, bathrooms and utility rooms. Typical exhaust rates are 50 cfm, 20 cfm and 25 cfm, respectively. The amount of supply air should equal the exhaust air, with most of the supply going to the main living area and lesser amounts to sleeping areas.

Adjustable dampers installed in the ducts make balancing relatively easy. A damper on the intake line ensures that the airflow rate can be adjusted to equal the exhaust air rate. Airflow rates can be compared visually with the aid of smoke pencils or tissue paper, but for accurate measurement, pressure gauges and charts are necessary. Fortunately, several HRV manufacturers now sell these specialized tools to aid in system balancing.

Putting an end to problems—It should be obvious by now that HRVs are not any more complex than dozens of other technological wonders we surround ourselves with these days, and much simpler than many. Why, then, have there been so many problems? One reason is that HRVs are still new to many builders. Even in areas (like Washington and Oregon) where building codes have been upgraded to encourage energy-efficient construction, specifications for heat-recovery ventilation are vague at best, and standards for operation and installation are just now being set up.

In the late 1970s and early 1980s, techniques for building tight, highly insulated houses ran far ahead of the technology for residential HRV units. News of fantastic energy savings was later blunted by reports of health hazards due to indoor air pollution. Builders and HRV suppliers were caught flat-footed, unprepared for the technical problems involved in properly designing hundreds of HRV systems, getting them properly installed, and keeping them serviced. Home owners weren't familiar with this new breed of appliance either, so they couldn't tell if their HRV system was performing correctly or not.

Thanks to the hundreds of test houses that have been built and monitored over the last few years in the U. S. and Canada, we now know a lot more about HRV sizing, system design and maintenance. The first-generation HRVs that fared so badly in the tests I mentioned earlier have largely been replaced by updated models that are more reliable. Still, the HRV today is somewhat like the car at the turn of the century—automobiles were being made and sold, but the support network (gas stations, mechanics, parts suppliers) was almost non-existent in many areas.

But things are getting better—HRV reliability is improving, incompetent manufacturers are going out of business, and the support network is growing. For example, Conservation Energy Systems (vanEE), one of the oldest HRV manufacturers, has developed the first HRV control that monitors air quality. Aircheck® samples household air and automatically runs the HRV at full speed if the concentration of certain indoor air pollutants exceeds a preset level. □

John R. Hughes is a freelance writer and a designer/consultant on energy-efficient housing, based in Edmonton, Alta.

Ridge-Vent Options

Two site-built alternatives to the standard aluminum fixture

Editor's note: Until a few years ago, roof venting was pretty much confined to open attics. The target then was heat buildup in the summer. Fixed-louver gable-end vents of wood or sheet metal, or mushroom (eyebrow) vents and turbine ventilators were used on the roof to get air flowing through the attic on days when there was a breeze. Moisture accumulation in the winter wasn't a serious problem in most climates, so venting a cathedral ceiling (where the finished ceiling is nailed directly to the rafters) was seldom even considered.

But with the demand for tighter construction, a growing awareness of the problems trapped moisture can cause, and the threat of ice-damming in heavy snow areas, roof venting has become an essential part of building. Model-code requirements call for a total Net Free Ventilation Area that's 1⁄150 of the total square footage of the ceiling below the roof. The Net Free Ventilation Area, or NFVA, is the total opening of the vent, usually measured in square inches, after the area taken up by screening or louvers has been subtracted from the total coverage of the vent. The 1⁄150 ratio can be reduced to 1⁄300 if a vapor barrier is used on the underside of the ceiling joists or rafters, or if the vented areas are split equally between low (soffit) vents and high (ridge, gable or roof) vents.

Venting-hardware manufacturers use a ventilation rate expressed in cubic feet per minute flow as a measure, and call for an even greater NFVA: .4 to 1.0 cfm per sq. ft. of floor area for winter moisture removal, and 1.5 to 2.0 cfm per sq. ft. of floor area for summer heat removal ("Attic Venting," p. 118). But many builders feel this much venting isn't necessary, given the widening use of vapor barriers and the leaky nature of even the tightest wood-frame house.

Our mail shows that our readers, both novice and professional builders, are trying to get a handle on when venting is necessary, how much should be supplied, and how best to build it in. The most effective way to vent a gable roof is to combine strip soffit vents and a continuous ridge vent. This system makes use of the fact that warm air rises, establishing a convection current from the eaveline to the ridge. It is also the only way to vent a shed or gable roof when the finished ceiling is nailed to the bottom of the rafters.

The standard fixture for the ridge is an extruded-aluminum vent that runs continuously along the peak. But a lot of our readers think manufactured ridge vents are irritatingly conspicuous as the crowning touch of a roof, especially when the vents are dented—a not infrequent occurrence considering the light-gauge metal (about .019 in.) from which they are made.

Two such critics are Doug Amsbary, a building contractor in Franconia, N. H., and Eric Rekdahl, an architect and builder in Berkeley, Calif. Amsbary remarked that he had seen too many vents bent and battered by snow shovels and chimney sweeps' ladders. He was looking for an alternative that was more attractive and durable for the custom houses he builds.

In the mild climate of California, where continuous ridge vents are less common, Rekdahl found himself needing to vent cathedral ceilings without using add-on hardware that would detract from his designs. Both of these builders have come up with alternative ridge vents made from standard lumber and roofing material, and have used them on houses they built last year. Their designs, quite different in style and structure, are detailed here. Like most site-built solutions to new problems, these vents are a first-generation response. Neither contains a full wind baffle, which can be important in high wind areas to prevent the convection currents from the soffit to the ridge from reversing, and to keep blowing snow and rain out. However, both Amsbary and Rekdahl are satisfied so far that insulation and sheathing are staying dry. There are lots of ways to solve venting problems; and these two designs will undoubtedly be refined as more are built. One reason for presenting them is to offer others a starting point for their own solutions. —Paul Spring

---

by Doug Amsbary

Manufactured aluminum ridge vents are easy enough to install, and they do their jobs well. No argument. But I take a lot of care with the custom homes I build, and I set about last summer to design a ridge vent that I could make on site that would be sturdier and more attractive. I started with blocks that are toenailed to the rafters at the peak, and incorporate a commercial aluminum strip vent 2¼ in. high. The rest is just sheathing, drip edge and asphalt shingles. The materials cost just a little more than an aluminum ridge vent (see the comparison chart on the next page), but there's quite a bit more labor in my design. After fabricating four of these site-built vents this summer, though, I decided it was worth the extra time.

These vents can be used on either new work or renovation. Make sure that there will be room (1½ in. to 2 in.) between the insulation and the sheathing for a continuous flow of air from the soffit vent at the bottom of each rafter bay to the ridge on top.

On a retrofit, the first task is to strip off the ridge shingles and the first few courses below on either side of the peak. Then, carefully lay out where the bottom edge of the blocks will be toenailed to the rafters (the blocks I have been using are 7¾ in. from the long point of the pitch cut at the ridge to the square cut on the downhill end). Add 1⁄16 in. to ⅛ in. to your calculations when cutting the sheathing back so that there is enough room to slide the vent strip into place. Take several measurements along the ridge in case the peak isn't straight. Now snap a line and make the cut on the sheathing. If the sheathing isn't plywood, small nailers might have to be added to support the sheathing boards where the vent terminates next to a rafter. On new construction, wait to sheathe the upper portion of the roof until you've installed the vent strip. Otherwise the sheathing will get in the way of nailing the bottom flange of the aluminum venting.

Ridge blocks—I use 2x softwood for the ridge blocks. They are nailed directly on top of the rafters, and must extend above the sheathing exactly 2¼ in. Before cutting all of the blocks, give the first pair a trial fit at several spots on the ridge so that you can adjust the plumb cut if necessary. Use fairly clear stock for the ridge blocks so that split-out is minimized. Make a few spares while you're at it. In fairly dense framing wood, predrilling may be necessary.

The blocks should be toenailed into the existing rafters. I use four nails for each block—one on each side of the block, one on the downhill end, and the last one driven down from the top of one block into the other. It is a good idea to install the end sets of blocks first and then string a line between them. I found that if the rafters didn't join the peak either at the same height or along a straight line laterally, the precut blocks tended to pick up the sweeps and dips. This makes it frustrating to install both the vent sheathing and the vent

Material cost breakdown	
Standard ridge vent	**$22.49**
10-ft. length of aluminum vent	18.00
1 connector	1.90
1 end cap	.80
10 oz. asphalt roof cement	1.79
Site-made ridge vent	**$26.09**
2 pcs. 10-ft. 1x10 #3 pine	6.64
Ridge blocks—12-ft. 2x3 spruce	1.92
2½ 8-ft. pieces #SV202 @$1.45	4.10
2 pcs. 5-in. galvanized drip edge	2.90
8d and 16d nails @ $.40 lb.	.20
1-in. galv. roofing nails @ $.65 lb.	.38
1 bundle ridge shingles	9.95

Based on suppliers' prices in 1983 for materials bought in quantity. Comparison based on 10 lin. ft.

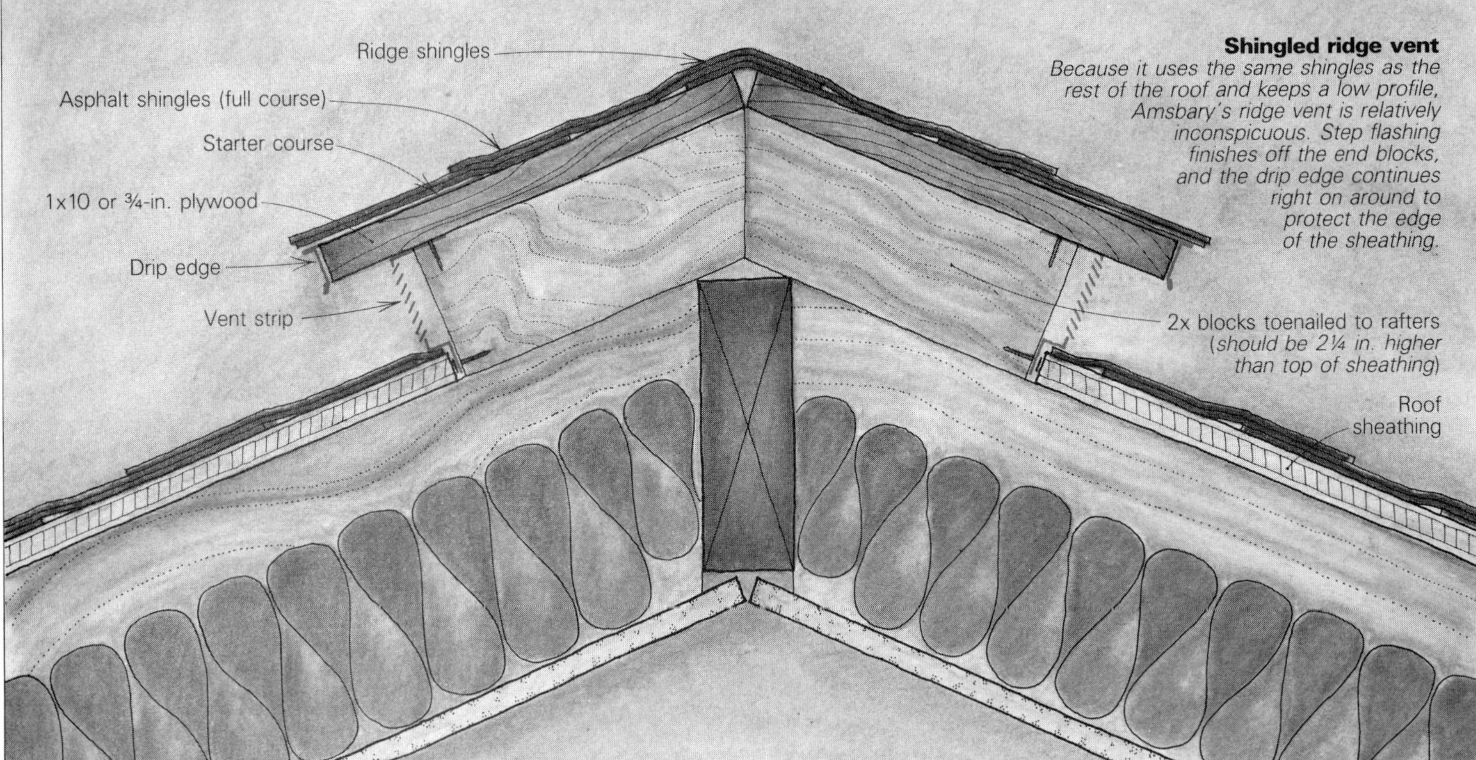

Shingled ridge vent
Because it uses the same shingles as the rest of the roof and keeps a low profile, Amsbary's ridge vent is relatively inconspicuous. Step flashing finishes off the end blocks, and the drip edge continues right on around to protect the edge of the sheathing.

strip itself. A bit of planing or ripping on the blocks may be necessary at this point if the ridge is a real roller coaster. In any remodeling, compromise is the key, and a string line will help you define that compromise here.

Vent strip—The aluminum vent strip I use—Model SV 202 by Air Vent Inc. (6907 N. Knoxville Ave., Peoria, Ill. 61614)—is intended for soffits, but with its low profile and small (⅛-in.) louver openings, it works well for the ridge. This vent strip comes only in a mill finish, but the pieces can be spray-painted to match the roof color.

You can use either staples or 4d box nails to secure the top and bottom flanges of the vent strip. On new construction you will be able to nail the bottom flange directly to the block ends, if you hold the sheathing back until you're finished. On a retrofit job, you will only be able to catch the edge of the roof sheathing by nailing from inside the vent between the blocks.

Vent strips come in 8-ft. lengths. Lapping the pieces isn't necessary, but do butt them over the nearest set of ridge blocks. Make sure your vent strip extends slightly (1/16 in.) beyond the end set of ridge blocks so the step flashing you'll be using as a finished end cap will fit tightly beside the vent strip.

Sheathing and roofing—The next step is sheathing the little gable you've created. You can use either ¾-in. plywood or 1x softwood solid stock. So far I've cut these pieces 10 in. wide to the long point of the peak, but 12 in. might help cut down on any possibility of driving rain working its way through the vent strip. It would also cut down on the number of times I accidentally kicked the vent strip after it was installed.

Whatever width sheathing you use, make sure that it overhangs the blocks by at least 1 in. around the entire perimeter of the vent. To protect the edge of the sheathing, install either galvanized or aluminum drip edge using ¾-in. roofing nails. This kind of attention to detail during construction will mean that the vent won't require any more maintenance than the roof it sits on. Next apply a starter course of asphalt shingles on both sides of the vent—use the top portion of the shingle ripped down to include the self-sticking asphalt adhesive. Then nail on the first (and only) course of shingles on each side of the vent, being careful to line up the cutouts with the pattern already established on the roof. Cap this peak with standard ridge shingles laid toward the prevailing wind, and you're finished with the top of the vent.

As you approach the bottom edge of the vent from each side of the main roof, be sure that you hold back the next-to-last course of shingles about an inch so that the last course can fit tightly beneath the ⅜-in. offset at the vent strip's bottom edge. These last shingles will have to be face-nailed. Use a spot of roof cement on the top of each exposed nail. Also run a small bead of plastic roof cement along the juncture between the vent strip and the short top course of shingles to keep everything dry underneath.

To seal off the exposed outside face of the last set of blocks, use step flashing, or make up the necessary pieces using colored coil stock, aluminum or copper. Cut this flashing so that it fits tightly against the underside of the ridge-vent sheathing. □

Illustrations: Vince Babak

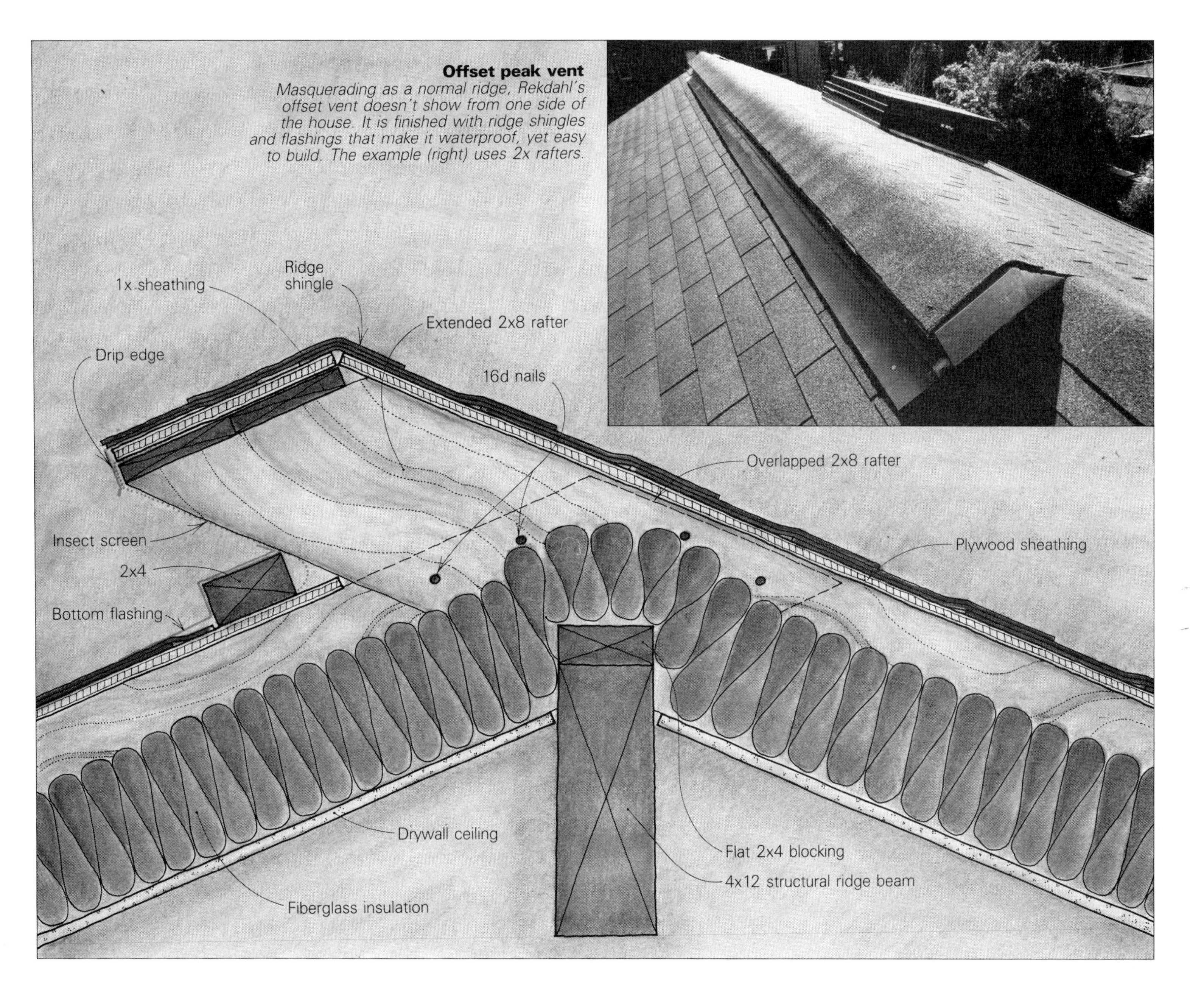

Offset peak vent

Masquerading as a normal ridge, Rekdahl's offset vent doesn't show from one side of the house. It is finished with ridge shingles and flashings that make it waterproof, yet easy to build. The example (right) uses 2x rafters.

by Eric Rekdahl

Last winter I was faced with designing a two-story addition whose roof was limited in height from the outset by the height of the existing one-story roof nearby. Even using a slab on grade for the addition's ground floor to stay as low as possible, I had to nail the gypboard ceiling in the addition's partial loft directly to the underside of the gable roof rafters to get the required head clearance. This plan left me no choice but to vent the roof from the soffit up through the ridge. I began experimenting at my drawing board with a way to incorporate a ridge vent without breaking the plane of the gable roof on at least the one side that faces the yard, and without adding a lot of conspicuous hardware. The solution I came up with creates a continuous ridge vent by lapping the rafters coming from each side, and letting one side run long beyond the normal roof peak. These long rafters are cut off at the same angle as the return side of the roof. Since the top edges of these rafters are higher than anything else on the roof, they create an offset ridge. With standard ridge shingles capping this peak, it looks like a normal gable from one side of the ridge. But on the other side, the extra rafter length creates a miniature clerestory, whose intervening spaces provide the venting.

A nice side benefit of the overlapping roof rafters is the increased resistance to structural separation. Each rafter has full bearing on top of the structural ridge beam, and by spiking them together at the top, they interlock nicely around the beam. To allow an unimpeded flow of air from the soffit vent at the bottom of the rafters to the ridge vent at the top, we nail 2x4 blocking flat on top of the ridge beam, and use metal crossbracing to hold the rafters in position at midspan.

The vent detailing is simple and inexpensive. First we nail 1x sheathing to the end of the extended rafter (this is the short return pitch of the offset peak). Once the screening is stapled down, this area is built up with ½-in. plywood. A 1½-in. by 2-in. drip edge is then nailed to the top of the plywood to protect its edges and the sheathing below it.

The screening is simply ¹⁄₁₆-in. insect screen (we used a dark-colored fiberglass screening) available off the roll from most hardware stores. Code in my area now requires ¼-in. insect screen to meet Net Free Ventilation Area needs, but this size mesh will keep only large insects out. We start the screening high on the 1x sheathing of the little return, and stretch it down across the opening and onto the slope of the main roof. We use a staple gun or hammer tacker, and tack it every few inches.

The bottom of the screened openings requires a flat 2x4 nailed to the slope of the lower, main roof and pushed tightly against the bottom edge of the extended rafters, as shown in the drawing. The undercourse of 15-lb. felt I use with asphalt shingles on top of the roof sheathing is wrapped up over this 2x4. It's capped with a simple flashing that begins at the top with a ¾-in. lip that sits in front of the screening. This prevents blowback—water that has fallen on the flashed surface being driven back toward the vent opening by the wind. The flashing then slopes down the face of the 2x4, folds over its bottom edge and lies flat on the asphalt shingles below. We had ours made up by a local sheet-metal shop in 10-ft. lengths. □